T0252505

# Structural Lightweight Aggregate Concrete

*To the many people who, in the past, worked in the Research Departments of the British Cement Association and, previously, the Cement and Concrete Association.*

# STRUCTURAL LIGHTWEIGHT AGGREGATE CONCRETE

Edited by

**JOHN L. CLARKE**
Chief Structural Engineer
British Cement Association
Crowthorne
Berkshire

**CRC Press**
Taylor & Francis Group
Boca Raton  London  New York

CRC Press is an imprint of the
Taylor & Francis Group, an **informa** business

A TAYLOR & FRANCIS BOOK

CRC Press
Taylor & Francis Group
6000 Broken Sound Parkway NW, Suite 300
Boca Raton, FL 33487-2742

First issued in paperback 2019

© 1993 by Taylor & Francis Group, LLC
CRC Press is an imprint of Taylor & Francis Group, an Informa business

No claim to original U.S. Government works

ISBN-13: 978-0-7514-0006-9 (hbk)
ISBN-13: 978-0-367-86344-9 (pbk)

This book contains information obtained from authentic and highly regarded sources.
Reasonable efforts have been made to publish reliable data and information, but the
author and publisher cannot assume responsibility for the validity of all materials or
the consequences of their use. The authors and publishers have attempted to trace
the copyright holders of all material reproduced in this publication and apologize to
copyright holders if permission to publish in this form has not been obtained. If any
copyright material has not been acknowledged please write and let us know so we
may rectify in any future reprint.

Except as permitted under U.S. Copyright Law, no part of this book may be reprinted,
reproduced, transmitted, or utilized in any form by any electronic, mechanical, or other
means, now known or hereafter invented, including photocopying, microfilming, and
recording, or in any information storage or retrieval system, without written
permission from the publishers.

For permission to photocopy or use material electronically from this work, please
access www.copyright.com (http://www.copyright.com/) or contact the Copyright
Clearance Center, Inc. (CCC), 222 Rosewood Drive, Danvers, MA 01923, 978-750-8400.
CCC is a not-for-profit organization that provides licenses and registration for a
variety of users. For organizations that have been granted a photocopy license by the
CCC, a separate system of payment has been arranged.

**Trademark Notice:** Product or corporate names may be trademarks or registered
trademarks, and are used only for identification and explanation without intent to infringe.

Typeset in 10/12 pt Times New Roman by Best-set Typesetter Ltd.,
Hong Kong

A catalogue record for this book is available from the British Library

Library of Congress Cataloging-in-Publication data available

Visit the Taylor & Francis Web site at
http://www.taylorandfrancis.com

and the CRC Press Web site at
http://www.crcpress.com

# Preface

Concrete is the most widely used building and construction material in the world. Natural lightweight aggregates have been used since Roman times and artificial lightweight aggregates have been available for over 70 years. However, lightweight aggregate concrete has only a very small share of the market. A number of notable structures have been built in the UK, Europe and the USA and yet many designers appear to ignore the material. Some reject it on the grounds of the basic cost, taking no account of the benefits that can result from its use, such as reduced member sizes, longer spans, improved fire resistance, smaller foundations and better thermal properties.

Lightweight aggregate concrete is covered, briefly, in most structural design codes but reference is generally made to specialist documents for more detailed information. This book aims to bring together all aspects of the material, considering the manufacture of the aggregates, mix design and construction, design requirements and specific applications in buildings, bridges and other structures. Information has been included not only from the UK but also from the rest of Europe, the USA and Japan. The authors of the various chapters all have extensive experience of lightweight aggregate concrete and are drawn from all branches of the industry.

This book is intended for all those who may be concerned with lightweight aggregate concrete, be they specifiers, materials suppliers, designers, contractors or the eventual owners of the building or structure. It is hoped that, by dealing with all the aspects, this book will help lightweight aggregate concrete to achieve its rightful place in construction.

J. L. C.

# Contributors

**Mr B. K. Bardhan-Roy**  Jan Bobrowski & Partners, Grosvenor House, Grosvenor Road, Twickenham, Middlesex, TW1 4AA, UK

**Mr J. L. Clarke**  British Cement Association, Century House, Telford Avenue, Crowthorne, Berkshire, RG11 6YS, UK

**Mrs D. Lazarus**  Ove Arup & Partners, 13 Fitzroy Street, London, W1P 6BQ, UK

**Mr J. H. J. Manhoudt**  BVN Raadgevend Ingenieursbureau B.V., Volmerlaan 20, 2288 GD Rijwijk (Z.H.), The Netherlands

**Mr J. B. Newman**  Concrete Structures Section, Civil Engineering Department, Imperial College, London, SW7 2AZ, UK

**Mr P. L. Owens**  Quality Ash Association, Rosebank, Donkey Lane, Tring, HP23 4DY, UK

**Mr R. N. W. Pankhurst** 4 Elmfield, Great Bookham, Surrey, KT23 3LQ, UK

# Contents

**1 Lightweight aggregates for structural concrete**    **1**
P. L. OWENS

1.1 Introduction, definitions and limitations    1
1.2 Lightweight aggregates suitable for use in structural concrete    2
1.3 Brief history of lightweight aggregate production    3
1.4 Manufacturing considerations for structural grades of lightweight
     aggregate    5
     1.4.1 The investment    5
     1.4.2 The resource materials    5
     1.4.3 The various processes of lightweight aggregate manufacture    6
     1.4.4 The techniques of production    6
1.5 Lightweight aggregates available in the UK    7
1.6 Production methods used for the various lightweight aggregates    7
     1.6.1 Foamed slag    7
     1.6.2 Leca and Fibo    8
     1.6.3 Lytag    10
     1.6.4 Pellite    12
     1.6.5 Granulex and Liapor    12
1.7 The future    14
1.8 Conclusions    17
References    17

**2 Properties of structural lightweight aggregate concrete**    **19**
J. B. NEWMAN

2.1 Introduction    19
2.2 Properties of lightweight aggregate for structural concrete    20
2.3 Properties of structural lightweight aggregate concrete    22
     2.3.1 Fresh concrete    22
     2.3.2 Density    23
     2.3.3 Strength    24
     2.3.4 Strength/density ratio    29
     2.3.5 Impact    31
     2.3.6 Deformation    31
     2.3.7 Bond and anchorage    33
     2.3.8 Fatigue    34
     2.3.9 Durability    34
     2.3.10 Thermal behaviour    37
     2.3.11 Acoustic behaviour    39
     2.3.12 Fire resistance    39
2.4 Experience in use    40
References    41

## 3 Design requirements 45

J. L. CLARKE

| | | |
|---|---|---|
| 3.1 | Provision for lightweight aggregate concrete in codes | 45 |
| | 3.1.1 Introduction | 45 |
| | 3.1.2 British codes | 46 |
| | 3.1.3 American codes | 48 |
| | 3.1.4 Norwegian code | 48 |
| | 3.1.5 European code | 48 |
| | 3.1.6 Australian code | 49 |
| | 3.1.7 Japanese specifications | 49 |
| 3.2 | Design requirements for reinforced concrete | 49 |
| | 3.2.1 Introduction | 49 |
| | 3.2.2 Definition of lightweight concrete | 50 |
| | 3.2.3 Limitations on compressive strength | 50 |
| | 3.2.4 Cover to reinforcement | 51 |
| | 3.2.5 Fire | 53 |
| | 3.2.6 Flexure | 55 |
| | 3.2.7 Shear resistance of beams | 55 |
| | 3.2.8 Torsion | 60 |
| | 3.2.9 Deflections | 61 |
| | 3.2.10 Shear of slabs | 63 |
| | 3.2.11 Columns | 63 |
| | 3.2.12 Walls | 64 |
| | 3.2.13 Detailing of reinforcement | 64 |
| 3.3 | Design requirements for prestressed concrete | 66 |
| | 3.3.1 Introduction | 66 |
| | 3.3.2 Cover to reinforcement for durability and fire | 66 |
| | 3.3.3 Service and transfer conditions | 67 |
| | 3.3.4 Shear of beams | 67 |
| | 3.3.5 Prestress losses | 68 |
| | 3.3.6 Transmission length | 69 |
| 3.4 | Thermal effects | 69 |
| | 3.4.1 Early thermal cracking during construction | 69 |
| | 3.4.2 Thermal movements in mature concrete | 70 |
| 3.5 | Overall design implications | 71 |
| | 3.5.1 Introduction | 71 |
| | 3.5.2 Cover to reinforcement | 72 |
| | 3.5.3 Flexure | 72 |
| | 3.5.4 Shear of beams | 72 |
| | 3.5.5 Shear of slabs | 72 |
| | 3.5.6 Deflections | 72 |
| | 3.5.7 Columns | 72 |
| | 3.5.8 Detailing | 73 |
| | 3.5.9 Prestressed concrete | 73 |
| References | | 73 |

## 4 Construction 75

R. N. W. PANKHURST

| | | |
|---|---|---|
| 4.1 | Introduction | 75 |
| | 4.1.1 Historical background | 75 |
| | 4.1.2 Lightweight aggregate in concrete | 76 |
| 4.2 | Supply of lightweight aggregate | 78 |
| | 4.2.1 Bulk density and moisture content | 78 |
| | 4.2.2 Controlling moisture content | 79 |

|       |       |                                                | |
|-------|-------|------------------------------------------------|-----|
| 4.3   | Mix designs |                                          | 80  |
|       | 4.3.1 | Introduction                                   | 80  |
|       | 4.3.2 | Lightweight fines                              | 81  |
|       | 4.3.3 | Pumped concrete                                | 81  |
|       | 4.3.4 | Mix designs for pumping                        | 82  |
| 4.4   | Batching |                                             | 83  |
|       | 4.4.1 | Aggregate proportion                           | 83  |
|       | 4.4.2 | Mixing                                         | 84  |
|       | 4.4.3 | Yield                                          | 85  |
| 4.5   | Pumping |                                              | 86  |
|       | 4.5.1 | Developments in pumping practice               | 86  |
|       | 4.5.2 | Pumping for high-rise buildings                | 88  |
|       | 4.5.3 | Canary Wharf trials and experience             | 89  |
|       | 4.5.4 | Recommendations for pumping                    | 91  |
| 4.6   | Placing, compaction and finishes |                     | 93  |
|       | 4.6.1 | Formed finishes                                | 93  |
|       | 4.6.2 | Floor slabs                                    | 93  |
|       | 4.6.3 | Unformed finishes                              | 94  |
|       | 4.6.4 | Power floating                                 | 94  |
|       | 4.6.5 | Computer floors                                | 95  |
|       | 4.6.6 | Weather                                        | 95  |
|       | 4.6.7 | Vacuum de-watering                             | 97  |
| 4.7   | Testing lightweight aggregate concrete |               | 97  |
|       | 4.7.1 | Strength                                       | 97  |
|       | 4.7.2 | Workability                                    | 97  |
|       | 4.7.3 | Testing for density                            | 98  |
|       | 4.7.4 | *In-situ* strength testing                     | 99  |
|       | 4.7.5 | Performance in fire                            | 99  |
|       | 4.7.6 | Fixing into lightweight aggregate concrete     | 100 |
|       | 4.7.7 | Making good lightweight aggregate concrete     | 100 |
|       | 4.7.8 | Productivity                                   | 100 |
| 4.8   | Economics |                                            | 101 |
| 4.9   | Conclusions |                                          | 103 |
|       | Appendix |                                             | 104 |

# 5 Lightweight concrete in buildings

**106**

D. LAZARUS

|       |       |                                                | |
|-------|-------|------------------------------------------------|-----|
| 5.1   | Introduction |                                         | 106 |
|       | 5.1.1 | General                                        | 106 |
|       | 5.1.2 | Historical perspective                         | 106 |
| 5.2   | Factors in the selection of lightweight aggregate concrete | | 107 |
|       | 5.2.1 | Introduction                                   | 107 |
|       | 5.2.2 | Durability                                     | 107 |
|       | 5.2.3 | Fire                                           | 108 |
|       | 5.2.4 | High-strength concrete                         | 109 |
|       | 5.2.5 | Placing lightweight aggregate concrete         | 110 |
|       | 5.2.6 | Slipforming                                    | 111 |
|       | 5.2.7 | Finishes                                       | 112 |
|       | 5.2.8 | Finishing                                      | 112 |
| 5.3   | Applications of lightweight aggregate concrete |       | 113 |
|       | 5.3.1 | *In-situ* concrete structures                  | 113 |
|       | 5.3.2 | Composite slabs with profiled metal decking    | 126 |
|       | 5.3.3 | Precast units                                  | 129 |
|       | 5.3.4 | Blockwork                                      | 136 |
|       | 5.3.5 | Refurbishment                                  | 141 |
| 5.4   | Economics of lightweight aggregate concrete in buildings | | 143 |

    5.4.1   Introduction                                          143
    5.4.2   The Concrete Society study                            143
    5.4.3   Other information                                     145
    5.4.4   The selection of lightweight aggregate concrete       148
    References                                                    148

6   **Lightweight concrete in bridges**                          **150**
    J. H. J. MANHOUDT

    6.1   Introduction                                            150
    6.2   Why use lightweight concrete in bridges?               150
    6.3   Types of aggregates used for bridges                    152
    6.4   Advantages and disadvantages                            153
    6.5   Recent research                                         155
    6.6   Recommendations for applications in bridges             156
    6.7   Examples of bridge structures                           157
          6.7.1   Koningspleijbrug, a bridge near Arnhem, the Netherlands   157
          6.7.2   Bridge at Redesdale, UK                         162
          6.7.3   Bridge at Ringway (ring road) near Ulft, the Netherlands  162
          6.7.4   Bridge over the river Sinigo at Avelengo (Bolzano), Italy  164
          6.7.5   The Friarton Bridge in Scotland                 165
          6.7.6   Refurbishment and upgrading                     166
    6.8   Summary and conclusions                                 166
    References                                                    166

7   **Lightweight concrete for special structures**              **168**
    B. K. BARDHAN-ROY

    7.1   Introduction                                            168
    7.2   Concrete quality                                        169
    7.3   Examples of applications                                170
          7.3.1   Marine and offshore structures                 170
          7.3.2   Onshore structures                              179
    References                                                    193

**Appendix 1**                                                   **195**

    Design of a road bridge using standard prestressed M beams in lightweight
    aggregate concrete

**Appendix 2**                                                   **219**

    Design of a cantilever roof beam of a grandstand structure using prestressed
    lightweight concrete

**Appendix 3**                                                   **229**

    Design of lightweight concrete prestressed double-T unit construction for 4
    hours' fire resistance

**Index**                                                        **237**

# 1 Lightweight aggregates for structural concrete
## P. L. OWENS

### 1.1 Introduction, definitions and limitations

Structural lightweight concrete is defined as having an oven-dry density of less than $2000 \, kg/m^3$ [1]. The aggregates used may be a combination of fractions of both lightweight coarse and fine materials or lightweight coarse material with an appropriate, natural fine aggregate. However, the general term lightweight concrete refers to any concrete produced to an oven-dry density of less than $2000 \, kg/m^3$. This can be achieved with most natural aggregates if the concrete is made in such a way that excess air is incorporated for example, as no fines or foamed concrete. These types of concrete are outside the scope of this book and are not considered further.

Any aggregate with a particle density of less than $2000 \, kg/m^3$ or a dry loose bulk density of less than $1200 \, kg/m^3$ is defined as lightweight [2]. However, this necessary dual qualification in definition highlights a practical difference from most other aggregates used in structural concrete where particle densities greater than $2000 \, kg/m^3$ are used. In the case of an appropriate lightweight aggregate the encapsulated pores within the structure of the particle have to be combined with both the interstitial voids and the surface vesicles. Nevertheless, these features in combination should not increase the density of the compacted concrete either by significant water permeation (absorption) or cement paste pervasion into the body of the aggregate particle when the aggregate is mixed into concrete.

The most appropriate method of assessing particle density in structural concrete is related to that of an aggregate with a 'reference' density of $2600 \, kg/m^3$, in that this constitutes the major difference in what is a singular property of any lightweight aggregate. The minerals comprising the structure of most aggregates, whether lightweight or not, have densities close to $2600 \, kg/m^3$, but it is the retention of **air** within the structure of an aggregate, when in use, that enables compacted structural concrete to be less than $2000 \, kg/m^3$. Therefore producers of lightweight concretes, when requiring to compare the properties of one aggregate to another, should not only judge relativity, but also compare any differences in mass of the various concrete constituents, water, cement, etc., so that comparisons between aggregates are made strictly from the same base.

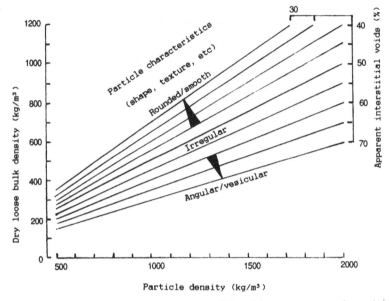

**Figure 1.1** Effect of apparent interstitial voids on lightweight aggregates, as determined by particle characteristics on the relationship of particle density to dry loose bulk density.

This aspect of apparent divergence between the various definitions of density, 'particle' versus 'dry loose bulk' is also concerned with the apparent percentage of interstitial voids and the way differently shaped particles of the same nominal size interact. The number of spherical particles, for a given volume, pack more naturally to a higher random density than do the same sized particles of either irregular or angular shape. To demonstrate this particular aspect of lightweight aggregate, Figure 1.1 illustrates the relationship between these two definitions of density, while incorporating the shape and texture of the particle, together with its apparent percentage void space. Thus those aggregates with the **lowest apparent percentage of void space** are those which give structural concrete densities significantly lower than 2000 kg/m$^3$.

## 1.2  Lightweight aggregates suitable for use in structural concrete

For structural concrete, the pragmatic requirements are generally that any lightweight aggregate is suitable that has a crushing strength sufficient to have reasonable resistance to fragmentation [3] while enabling concrete strengths in excess of 20 N/mm$^2$ to be developed and to produce a finished concrete in the dry density range 1500–2000 kg/m$^3$. Essentially, this means that where the concrete uses fine aggregates from natural sources, the particle density of the coarse aggregate in the 'compacted' concrete needs to be not greater than about $650\,kg/m^3$ at the lower end, nor more than about $1850\,kg/m^3$ at the higher end of the scale, because it is the degree of 'lightness' of the coarse aggregate that mostly influences the density of the finished oven-dry concrete.

Traditionally, producers and manufacturers of lightweight aggregates have been constrained by the following:

1. If from a natural source such as pumice, scoria, etc., the aggregate is what it is, so it has limited applications.
2. If manufactured, there are not only the limitations of the raw material and the method of processing, but also the main consideration, **the requirements of the market**. The market may be for an aggregate with good thermal insulation and not for structural applications. Added to this is the availability of the mineral resource, as the lower the finished aggregate particle density, the less will be the rate at which the resource will be depleted for the same volume of production. This creates an obvious conflict of interest: does a producer develop a lightweight aggregate for a market which depletes the resource at a greater rate, or for a speculative market with the possibilty of a smaller return on capital?

As a consequence of this dilemma, more lightweight aggregate producers have suffered failure than in any other segment of the aggregate industry because of conflicts in marketing strategy. However, it seems that most specifiers are limited in their appreciation of the advantages of lower density concrete in reinforced concrete construction, as the use of lightweight concrete has not been generally optimised. This is probably the result not only of inadequate design codes, which are more stringent than necessary as construction is a very 'conservative' and a traditional industry, but also of unfounded prejudice based on inadequate information at the design stage. Therefore, to overcome uncertainty requires demonstrable proof of long-term durability and stability of any concrete type, as well as the design information having to be presented in such a way that the advantages are immediately obvious.

## 1.3 Brief history of lightweight aggregate production

The history of lightweight aggregate production from natural sources dates back to pre-Roman times and continues today with volcanic porous rocks, but the sources are limited to regions of volcanic activity.

From around the end of the 19th century, with the development of reinforced concrete, and owing to the rarity of natural porous aggregate deposits and their non-existence in most developed countries, research for the manufacture of 'artificial' aggregates commenced. In Europe in the early part of the 20th century, development concentrated on foaming blastfurnace slags, as 'iron' production was basic to the industrial infrastructure. However, it was not until the early 1970s that significant developments in pelletising and expanding blastfurnace slags took place, so that today a slag-based aggregate with a smoother non-vesicular surface, more adaptable for structural concrete, is produced.

In comparison to slag, it was not until about 1913 that research in the USA revealed that certain clays and shales expanded when fired. This developed about 1917 in Kansas City, Missouri, to the production in a rotary kiln of a patented expanded aggregate known as Haydite which was used in the construction of the USS *Selma*, an ocean-going ship launched in 1919 [4]. There followed in the USA the development of a series of aggregates known as Gravelite, Terlite, Rocklite, etc. In Europe, however, it was not until 1931 that the manufacture of LECA (lightweight expanded clay aggregate) commenced in Denmark. Therafter developments quickly spread to Germany, Holland and the UK [5]. Variations to the principles of expanding suitable argillaceous materials, such as the geologically older forms of clay – e.g. shale and slate – have all been undertaken in the UK since the 1950s and have been variously known as Aglite, Brag, Russlite and Solite. All these companies have failed for one reason or another, mainly because of inaccessibility of the market, non-homogeneity of the feed stock, emissions and high cost of production, **but never on the technical performance of the aggregate in concrete**.

In the 1950s the Building Research Establishment (BRE) developed the technology for the production of a high-quality lightweight aggregate based on pelletised pulverised-fuel ash (PFA), an ash resulting from burning, generally in power stations, pulverised bituminous or hard coals [6]. PFA is the residue of contaminants in hard coals that are present as the result of erosion of natural minerals which sedimented into the coal measures as they formed. Thus there is a connection between PFA and other argillaceous minerals, except that most of the PFA has already been subjected to temperatures in excess of 1250°C, which vitrifies and bloats some of the larger particles, known as cenospheres. Two construction companies became involved in the exploitation of PFA and the BRE know-how. Cementation Ltd worked at Battersea Power Station, with two shaft kilns which failed operationally. John Laing & Co. Ltd set up at Northfleet, with a sinter strand which has evolved in the UK as the most successful method of producing structural grades of lightweight aggregate from PFA under the trade name of Lytag. Lytag has limited ability to reduce the density of fresh concrete much below 1750 kg/m$^3$ (when using Lytag fines) and can, with natural fine aggregate, produce a density of fresh concrete at about 1950 kg/m$^3$.

Where there is a requirement to produce high-strength concrete at densities lower than, say, 1850 kg/m$^3$, the manufacturers of Germany's expanded shale aggregate Liapor have, from about the early 1970s, been able to produce aggregates to within a desired particle density between 800 and 1700 kg/m$^3$. This highlights one of the most significant advances in lightweight aggregate manufacture as the **particle can be designed to suit the concrete density requirements**, thereby giving greater versatility to the application of this particular aggregate.

There are a number of other developments taking place in lightweight aggregate manufacture, such as the production of hybrids using pulverised-fuel ash and suitable argillaceous materials (clay, shales and slate) There are also what are euphemistically called 'cold' bonded aggregates which are mixtures of PFA with lime or Portland cement. These latter aggregates, although manufacture is now becoming more successful, have applications more appropriate to the production of masonry than to reinforced concrete.

## 1.4 Manufacturing considerations for structural grades of lightweight aggregate

### 1.4.1 The investment

For any lightweight aggregate the investment in manufacturing plant is considerable, for not only does there have to be sufficient and appropriate resource material available, but also there has to be a market. In the USA, most cities are based on the principle of high-rise development, with the inevitable use of lightweight aggregate. Meanwhile Europe, with its more historic traditions and older infrastructure, has lagged behind. Now more consideration is being given to the conservation of land-based resources such that the developers and promoters of various schemes find it more difficult to obtain permission, not only for mineral extraction, but also to acquire new sites for high-rise structures.

### 1.4.2 The resource materials

The most important asset for any lightweight aggregate manufacturer is to have sufficient raw material in a form and state ready for immediate use, which means that manufacturers using either PFA or molten slag have immediate advantages over other resource minerals. The limitations, however, are that the process of sintering PFA fuses the PFA particles in such a way that it densifies the aggregate, while to 'entrain' air into molten slag ultimately means that the particle density is limited to about 1750 kg/m$^3$.

Alternatively, an aggregate based on argillaceous materials such as clay, shale or slate can have its density varied by the manufacturing technique. In the case of the lighter aggregates they use less raw material, so for the stronger and denser aggregates the resource is depleted at a greater rate. One solution [7] is to make an aggregate based on PFA mixed with a minor proportion of clay, say in this case between 15 and 30%, which can be heat expanded to make an aggregate of the required particle density, say $1350 \pm 50 \, \text{kg/m}^3$.

### 1.4.3   The various processes of lightweight aggregate manufacture

Most manufacturing processes for lightweight aggregates, with the exception of processes using blastfurnace slag, have been limited to the use of either a sinter strand or a rotary kiln. In instances where the fresh pellets before firing are in a suitable form, the sinter strand is preferred. Where the form of the fresh pellet is cohesive and its shape can be retained, the rotary kiln produces the most rounded particle with the most impermeable surface.

### 1.4.4   The techniques of production

The various production techniques rely either on agglomeration or expansion (bloating). Agglomeration takes place when some of the materials melt at temperatures above 1100°C and the particles that make up the finished aggregate are bonded together by fusion. Alternatively, expansion develops when either steam is generated, as in the case of molten slag, or a suitable mineral (clay, shale or slate) is heated to fusion temperature, at which point pyroplasticity occurs simultaneously with the formation of gas which bloats the aggregate.

When argillaceous materials are heated by firing to achieve appropriate expansion, the resource mineral should contain sufficient gas-producing constituents and reach pyroplasticity at the point of incipient gas formation. Gas can be developed by a number of different reactions, either singularly or in combination, from the following:

(a) volatilisation of sulphides from about 400°C
(b) decomposition of the water of crystallisation from clay minerals at approximately 600°C
(c) combustion of carbon-based compounds from approximately 700°C
(d) decarbonation of carbonates from approximately 850°C
(e) reaction of $Fe_2O_3$, causing the liberation of oxygen from about 1100°C.

Most argillaceous materials that are suitable become pyroplastic at between 1100 and 1300°C. However , depending on the actual source of the material and its chemical composition, the temperature at which bloating for each material becomes effective is within a relatively small range, usually about ±25°C. At this point the bloated material has to be removed immediately from the firing zone and cooled quickly to freeze the particle at that degree of bloat, otherwise it will continue to expand. When ultimately the thickness of the pore wall becomes too thin, there is insufficient resistance to fragmentation and the particle will not be sufficiently strong to resist fragmentation for structural concrete.

A principle of success with all lightweight aggregate manufacture is homogeneity of the raw material source, as variability inevitably causes fluctuations in manufacture and the finished product. To emphasise this, it can be demonstrated how present-day manufacturers have had to go to considerable lengths to ensure that homogeneity of the raw material is obtained.

## 1.5 Lightweight aggregates available in the UK

The lightweight aggregates available in the UK, and currently listed in 1993, are given in Table 1.1.

**Table 1.1** Lightweight aggregates available in the UK (1993)

| Aggregate proprietary name | Type | Manufacturing process | Shape/texture | Dry loose bulk density (kg/m³) (typical) | Concrete strength (N/mm²) (typical) |
|---|---|---|---|---|---|
| Foamed slag | Foamed slag | Foaming bed | Angular/vesicular | 750 | <40 |
| Leca/Fibo | Expanded clay | Rotary kiln | Rounded/smooth | 425 | <30 |
| Lytag | Sintered PFA | Sinter strand | Rounded/fine | 825 | >40 |
| Pellite | Blastfurnace slag | Pelletisation | Irregular/smooth | 900 | >40 |
| Granulex | Expanded slate | Rotary kiln | Irregular/rough | 700 | >40 |
| Liapor | Expanded shale | Rotary kiln | Rounded/smooth: fine | 650 | >40 |

## 1.6 Production methods used for the various lightweight aggregates

### 1.6.1 Foamed slag

In this process, produced in the UK, molten blastfurnace slag at more than 1350°C is poured onto a foaming bed consisting of a large number of water jets set in a concrete base. The water immediately converts to steam on contact with the molten slag and penetrates into the body of the material, at which point the steam becomes superheated. Owing to the rapid expansion that then takes place, the slag foams to form a cellular structure. Alternative methods of expansion include spraying water onto the molten material when it is being tapped from the blastfurnace so that the material is cooled rapidly, with steam becoming entrapped within the structure of the particle. At the completion of foaming the slag is removed and stockpiled, from where it is subsequently crushed and graded to size (Figure 1.2). The aggregate produced is very angular with an open vesicular texture.

### 1.6.2  Leca and Fibo

Leca (produced in the UK) and Fibo (produced in Scandinavia) are expanded clay aggregates manufactured in a rotary kiln which consists of a long, large-diameter steel cylinder inclined at an angle of about 5° to the horizontal. The kiln is lined internally in the firing zone with refractory bricks which, as the kiln rotates, become heated to the required temperature and 'roast' the clay pellets for the required degree of expansion to occur. The length and configuration of the kiln depends in part on the composition of the clay and length of time it takes to 'condition' the clay pellet in the pre-heater to reach a temperature of about 650°C to avoid it shattering before becoming pyroplastic.

The clay is dug and, to eliminate natural variability, is usually deposited by layering into a covered stockpile with a spreader before it is removed from the stockpile by scalping with a bucket conveyor. In the process, a high degree of blending is achieved. The clay is prepared by mixing thoroughly to a suitable consistency before pelletisation. The prepared raw material – and that obviously means producing smaller sized pellets than are required for the finished product (as their volume can be increased three to six times) – is fed into the kiln. This can be in three segments, the higher end for drying and pre-heating, while at the lower end firing and then cooling takes place. During the progress of the prepared material through the kiln, the temperature of the clay pellets gradually rises until expansion actually occurs. The expanded product is discharged from the firing zone as soon as possible for cooling to freeze the particles at the required degree of expansion. Cooling takes place either in a rotary cooler or fluidised bed heat-exchanger. The finished product is graded and, if necessary, crushed to particle sizes less than 16 mm. While the particle density can be varied depending on the range of temperatures at which expansion takes place, the mean expansion temperature is about 1200 ± 50°C. This varies for different clays but most manufacturers are limited to expandable clays which have a confined range of bloating

(a)

(b)

Figure 1.2 (a) Typical Leca particles, 12 mm size. (b) Sectioned Leca particle about 30× enlarged.

temperatures. In some cases, the range is less than 25°C between non- and full expansion. In instances such as this, the scope is limited for the manufacture of intermediate density grades. However, manufacturers of such aggregates, by requiring to optimise their production, usually have preferences for lower particle density aggregates, in the range 400–800 kg/m$^3$, as this tends to conserve the resource which gets depleted at a greater rate at higher particle densities. Thus there is a greater attraction to produce aggregates for thermal insulation than for structural applications. However, for the lower density structural concrete, i.e. 1300–1600 kg/m$^3$, these aggregates are the most suitable. As shown in Figure 1.2, the surface texture is closed and smooth with a 'honeycombed' or foamed internal structure, where the pores are not interconnected.

### 1.6.3 Lytag

Lytag is produced in the UK from pulverised-fuel ash (PFA). Large quantities of suitable PFA are produced in the UK as a powdered by-product of pulverised-fuel (bituminous coal) operated furnaces of power stations. Suitable PFA, usually of less than 8% loss on ignition, which is mostly unburnt carbon in the form of coke, is first homogenised in bulk in its powder form. Once homogenised, it is then conditioned through a continuous mixer with about 12–15% water and, as necessary, an amount of fine coal is added to make up the fuel content to between 8 and 12% of the dry mass of the pellet to enable it to be fired. This conditioned mixture of PFA is then fed at a controlled rate onto inclined and revolving nodulising dishes. The inclination, the revolving speed, together with the rate at which the conditioned PFA is fed onto the dish, as well as some additional water added as a fine spray, controls the size and degree of compaction of the green pellets, which self-discharge from the pelletiser, when they become generally about 12–14 mm in size. Without any further treatment these green pellets are conveyed to the sinter strand where they are fed by spreading to form an open-textured and permeable bed to the width and depth of the continuously moving grate. This is in the form of a continuously moving conveyor belt, comprising a series of segmented and jointed grates through which combustion air can be drawn to fire the bed of nodules, as well as permitting exhaust for the gases of combustion. Once prepared on the bed, the strand immediately carries the pellets under the ignition hood that fires the intermixed 'fuel'. The chemical composition of PFA resembles that of clay, but unlike clay, as the PFA has already been fired, no pre-drying or pre-heating of the pellets is necessary, as the pellet is able to expire the water as vapour and combustion gases without incurring damage. Once ignited at about 1100°C, and as the bed moves forward, air for combustion is drawn by suction fans beneath the grate, and without the PFA particles becoming fully molten, bonding or coagulation of the PFA particles within the pellet is achieved. Controls for producing the correct amount of coagulation within the pellets are obtained by being able to vary both the speed of the strand and the amount of air drawn through the bed.

The finished product is formed into a block of hard brick-like spherical nodules, lightly bonded by fusion at their points of contact. As the sinter strand reaches the end of its travel and commences its return to the feeding station, a large segment of the finished product forming the bed is

(a)

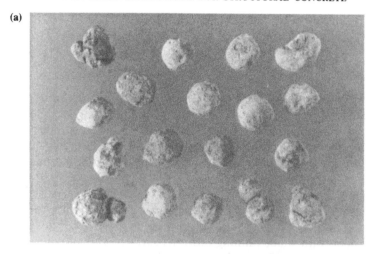

(b)

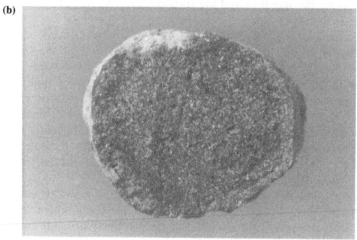

**Figure 1.3** (a) Typical Lytag particles, 12 mm size. (b) Sectioned Lytag particle about 30× enlarged.

discharged into a breaker. This has the ability to part pellets that are inevitably only lightly bonded together before the finished aggregate is graded.

The surface and internal structure of the finished pellet (Figure 1.3), while closed, is nevertheless sufficiently open textured to have encapsulated interstices between the coagulated PFA particles. While these interstices, although minute, are water permeable, they do eventually 'breathe' sufficiently, by emission, to allow any moisture to evaporate, even when encased in concrete.

### 1.6.4  Pellite

Pellite (produced in the UK) is a pelletised expanded blastfurnace slag. The process of slag pelletisation was developed in the early 1970s in order to overcome environmental problems associated with the production of foamed blastfurnace slag on open foaming beds or pits. Not only does the slag pelletisation process overcome these problems but it also produces an aggregate with a closed surface.

To manufacture this pelletised aggregate, liquid blastfurnace slag at a temperature of about 1400°C passes through a refractory orifice 'block' to control the rate of flow. It is then allowed to flow onto an inclined vibrating plate with water running down its surface. The vibration of the plate breaks up the slag flow and the trapped water immediately vaporises and expands the slag.

At the lower end of the vibrating plate, water is sprayed onto the surface of the slag. This enables gas bubbles to form in the body of the slag, creating further expansion while also chilling its surface. At the bottom of the vibrating plate the expanded globules of semi-molten slag are discharged onto a horizontally rotating drum fitted with fins which project the material through a water mist. The trajectory of material is such that the slag forms rounded pellets which are sufficiently chilled to avoid agglomeration when they come to rest. After pelletisation is complete the material is removed, allowed to drain and, finally, screened into the requisite size.

As Figure 1.4 shows, the nature of the manufacturing process produces a finished product that comprises semi-rounded pellets with a smooth surface encasing a glassy matrix and a discrete cellular structure, which is essentially non-absorbent.

### 1.6.5  Granulex and Liapor

Two other sources of European lightweight aggregate, besides Fibo, are available: Granulex and Liapor. They are not made in the UK but are, however, suitable for the production of high-strength or prestressed concrete. Both aggregates are produced by bloating argillaceous minerals in a rotary kiln.

**Granulex** is produced from slate in north-western France, just north of Le Mans. The product is very similar to the now defunct Solite, produced in north Wales. The slate is first reduced to about 12–15 mm, and is then fed into a three-stage kiln consisting of a pre-heater, followed by a firing or expanding section before being discharged into a cooler. The firing or

(a)

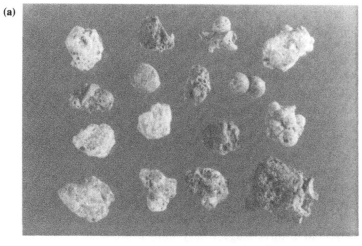

(b)

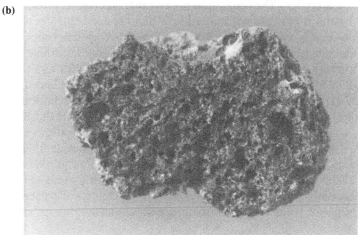

**Figure 1.4** (a) Typical Pellite particles, 12 mm size. (b) Sectioned Pellite particle about 30× enlarged.

bloating temperature is about 1150°C when the laminated platelets of slate become pyroplastic and the gases released cause the particles of aggregate to expand to form an almost cubic particle shape, taking the original particle density of the slate from about 2700 kg/m$^3$ to a pre-determined density anywhere between 650 and 1250 kg/m$^3$ depending on the required amount of bloating, although this has to be very carefully controlled. The finished product is crushed and graded up to 25 mm maximum size (Figure 1.5) and the surface texture of the finished particle is coarse and rough. However, the surface is sufficiently closed, owing to its vitrified nature, such that internally it has an extremely low water absorption.

**Liapor** is produced from shale in Bavarian Germany between Nurenberg and Munich. The shale, a low moisture content soft rock, is quarried and transferred by open tipper truck some 6 km to the processing plant, where it is reduced by primary crushers before being dried and milled into a powder, generally of less than 250 µm. It is then homogenised and stored ready for pelletisation. This process is similar to that used for making Lytag except that no fuel is added. However, after the pellets have been produced to the appropriate size, depending on the expansion required, they are compacted and coated with finely powdered limestone. The resultant pellets are spherical with a very high 'green' strength. They are then conveyed to a three-stage rotary kiln, a pre-heater, an expander, followed by a cooler. They are unlike any other aggregate produced by the use of natural argillaceous material, in that the feed stock is reduced to a powder and is then reconstituted into a pellet of predetermined size, and the amount of expansion can be controlled to give a particle of the required density. The technique used for producing the different particle densities is the control of firing temperature and the rotating speed of the firing kiln. The coating of limestone given at the time of finishing the green pellet increases the amount of surface vitrification to the finished product, which is very impermeable. Of all the aggregates currently available, Liapor gives the greatest versatility to the designer for pre-selecting an appropriate concrete density to suit the requirement. As Figure 1.5 shows, while the particle shape and surface texture of Liapor remain essentially the same, internally the porosity can be varied according to the amount of bloat required for the specified density.

## 1.7  The future

Research on structural concrete has established that appropriate light-weight aggregates **do not have the deficiencies in performance compared to concrete made with natural aggregates**. As the need increases for alternatives to aggregates from natural sources, there are opportunities for capitalising on the resources that exist, for which there is currently no accepted use in their present form. For instance, it is estimated that there are some 150 million tonnes of 'stockpiled' PFA in the UK. On the Continent and in other EC countries such a practice is now being penalised by environmental protection legislation by taxation of the power companies. Although Lytag has provided a basis for the exploitation of PFA,

(a)

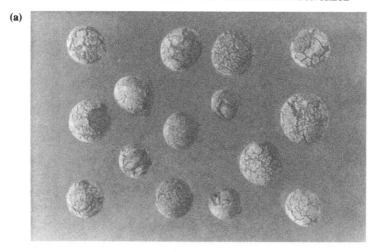

(b)

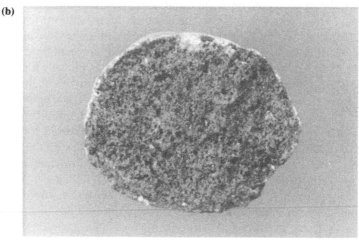

**Figure 1.5** (a) Typical Liapor particles, 12 mm size. (b) Sectioned Liapor particle about 30× enlarged.

the finished aggregate has a single density and, as such, has been limited in its ability for wider applications in structural concrete.

This, together with the limitations on the amount of excessive residual fuel that the PFA sometimes contains, restricts full capitalisation of the technology of PFA conversion to a structural aggregate. Alternatively, expanded clays have their own limitations, mainly because of restrictions placed on the availability of suitable material. In 1992 the one source of expanded clay aggregate produced in the UK has been threatened with closure as extension of the clay pit workings is being prevented by restricting planning permission to accessible resources. Therefore, Leca UK, quite naturally, needs to make the best use of its remaining available reserves by bloating the existing clay resource to the maximum, which thereby restricts the product's appropriateness for application in structural concrete.

There are, however, other alternatives by combining the technologies used in the manufacture of both Leca and Lytag to produce aggregates identical in performance to Liapor.

UK Patent No. 2218412 [7] exploits a suitable bloatable clay by using it to bind and pelletise PFA. The amount of clay used can be as low as 10% but can be increased to 80% by mass, if particle densities lower than $650 \, kg/m^3$ are required. The loss on ignition in the PFA is less restrictive as up to 20% can be accommodated. The advantages in manufacture are that the aggregate can be fired in a rotary kiln **without requiring a pre-heater** and the cooler can be a fluidised bed heat exchanger, both of which reduce the size of the plant to the essential rotary kiln for firing the product. Densities of the finished product can be more easily varied as the range of firing temperatures is increased beyond the restriction of the nominal $\pm 25°C$ for bloating while being able to use PFAs with much greater loss on ignition. The aggregates produced from this technology can have their particle densities varied in the range $400-1800 \, kg/m^3$. The particle is spherical with a low water permeability and smooth surface texture which, owing to the technology, can be manufactured for a full range of structural concrete (high and normal strength) in the range $1600-1750 \, kg/m^3$, as well as for aggregates with very low densities for thermal applications.

Another opportunity, but this time with the sinter strand, can use any argillaceous material – for instance, clay or the considerable stockpiles of 'soft' slate that was produced as waste in north Wales – while the spoil and tailings produced from mining coal are, in theory, potential raw materials for lightweight aggregate production. However, the volatiles in mine tailings, etc., present a considerable problem, while clay, shales and slate containing low amounts of volatile products are most suitable. However, for lightweight aggregate products to be successful in the future, pulverisation and reconstitution of the shale or slate into pellets, as done with Liapor, would have to be part of the process. UK Patent Application No. 92.25400.2 [8] proposes mixing a selected amount of between 10 and 50% of suitably prepared bloatable argillaceous material for introduction into a pelletisation process such as used for Lytag. This would not only lower the density of Lytag, but would also reduce its water absorption. While producing such an aggregate on the sinter strand the process would result in a less rounded particle than Lytag, but with the advantage of a closed surface, which, although being rough, would have an internal structure of greater porosity, with closed pores to reduce water absorption, and with characteristics similar to expanded slate (Granulex).

For increasing the suitability of mine 'tailings' as an aggregate feed stock, the removal of volatiles such as naphtha, sulphur and the bituminous products would seem an essential prerequisite, as well as pulverisation and homogenisation of the resultant 'clean' material before pelletisation.

## 1.8 Conclusions

The technology is available for successful lightweight aggregate manufacture in the UK, but there has to be considerable investment in process control of the feed material before any new generation of lightweight aggregates becomes available in addition to those that already exist. Solite, manufactured from the waste slate in north Wales, Brag manufactured from minestone in east Kent, as were Aglite in Derbyshire and Russlite in Scotland, all failed, not because of technical deficiencies of the aggregate itself, but because of difficulties with preparation of an inconsistent raw material and its process control. While the process for aggregate expansion was adequate in every case, it was the processing of the raw material itself which gave variability in process control. Alternatively, Pellite is produced from slag that is highly controlled as a result of iron production. Both Leca and Lytag have succeeded as in both processes considerable care is taken to remove variability from the raw materials to enable mass production to continue with as little 'manual' interference as possible.

The other problem with all lightweight aggregate manufacture is to match demand with supply, but significant investment will not be made into suitable lightweight aggregate production until it becomes a more common practice to use lightweight concrete for all structural purposes below, as well as above, ground level.

As the construction of USS *Selma* demonstrated almost 75 years ago, lightweight concrete was not only watertight but was also subsequently found to have considerable greater thermal strain capacity [9]. One Shell Plaza in Houston [10] used lightweight concrete in the foundations and basement, which proves that engineers can not only utilise the advantages of lightweight concrete, but can also make it commercially advantageous for owners of all types of structures.

## References

1. ENV 1992 – Euro Code No 2: Design of Concrete Structures Parts 1–4; The use of lightweight aggregate concretes with closed structures; Final Draft; October 1992.

2. BSI Document 92/17688: European Draft Standard Specification for Lightweight Aggregates. CEN/TC154/SC5, Sub-Committee Lightweight Aggregates, October 1992.
3. BSI Document 92/87196: European Draft Standard method of test for Crushing Strength of Lightweight Aggregates. CEN/TC154/SC5, Sub-Committee Lightweight Aggregates; Document N121; November 1992.
4. *Structural Lightweight Aggregate Concrete for Marine and Off-shore Applications*, Report of a Concrete Society Working Party, Concrete Society Technical Report No. 16, May 1978.
5. Rudnai, G., *Lightweight Concretes*, Akademiar Kiado Publishing House of the Hungarian Academy of Sciences, Budapest, 1963.
6. Cripwell, J. B., What is PFA?, *Concrete*, **26** (No. 3; May/June 1992), 11–13.
7. British Patent Specification No. 2218412: *Lightweight Aggregates*, Granted 22 April 1992.
8. British Patent Specification, Application No. 92.25400.2: *Lightweight Aggregates made with Pulverised-Fuel Ash*, Application date 4 December 1992.
9. Harrison, T. A., *Early Age Thermal Crack Control in Concrete*, revised edition, CIRIA Report 91, 1992.
10. Khan, F. R., Lightweight concrete for total design of One Shell Plaza, *ACI Special Publication SP29*, Lightweight Concrete Paper SP29–1, 1969.

# 2 Properties of structural lightweight aggregate concrete

J.B. NEWMAN

## 2.1 Introduction

Lightweight concretes can be produced with a density range of approximately 300–2000 kg/m$^3$, corresponding cube strengths from approximately 1 to over 60 N/mm$^2$ and thermal conductivities of 0.2 to 1.0 W/m K. These values can be compared with those for normal weight concrete of approximately 2100–2500 kg/m$^3$, 15 to greater than 100 N/mm$^2$ and 1.6–1.9 W/m K.

The properties of lightweight concrete can be exploited in a number of ways from its use as a primarily structural material to its incorporation in structures for the enhancement of thermal insulation. This variety of purpose is recognised by RILEM/CEB who have proposed the classification given in Table 2.1 [1].

This chapter discusses only those concretes within Class I (i.e. structural lightweight concrete).

The principal techniques used for producing structural lightweight concrete can be summarised as follows:

1. Omitting the finer fraction of normal weight aggregate to create air-filled voids using a process pioneered by Geo. Wimpey in the UK (1924). The resulting material is known as 'no-fines concrete'.
2. Including bubbles of gas in a cement paste or mortar matrix to form a cellular structure containing approx 30–50% voids ('aerated concrete').
3. Replacing, either wholly or partially, natural aggregates in a concrete mix with aggregates containing a large proportion of voids. Such concretes are usually referred to as lightweight aggregate concretes.

This chapter is concerned only with the latter group of concretes and, in particular, those made with lightweight aggregates within a Portland cement-based matrix (i.e. closed structure lightweight aggregate concretes).

Although structural lightweight concrete is usually defined as a concrete with an oven-dry density of no greater than 2000 kg/m$^3$ [2–5] there are variations in certain parts of the world. For example, in Australia [6] structural lightweight concrete is considered to be a concrete made with lightweight coarse aggregate and normal weight fines resulting in a

**Table 2.1** Classification of lightweight concretes

| Property | Class and type | | |
|---|---|---|---|
| | I Structural | II Structural/ insulating | III Insulating |
| Compressive strength (N/mm$^2$) | >15.0 | >3.5 | >0.5 |
| Coefficient of thermal condition (W/m K) | N/A | <0.75 | <0.30 |
| Approx. density range (kg/m$^3$) | 1600–2000 | <1600 | ≪1450 |

saturated surface-dry density of not less than 1800 kg/m$^3$. In Norway [7] any combination of any types of aggregate can be used for structural concrete provided the resulting concrete (a) has an oven-dry density of 1200–2200 kg/m$^3$ and (b) a strength grade of no greater than 85 N/mm$^2$ if the mix contains lightweight aggregate. In the USA [8] structural lightweight aggregate concrete is considered to be concrete with an air-dry density of less than 1810 kg/m$^3$. Two main classes of lightweight concrete are considered, namely, concrete made with lightweight coarse and fine materials and that made with lightweight coarse aggregate and natural fines. Interpolation between these classes is permitted. Perhaps the most radical approach is that adopted by the Japanese [9, 10] who refer to lightweight concrete but do not specify any density values and properties are only provided for concrete made with lightweight coarse and fine aggregates.

In Russia and the CIS lightweight concrete is defined in terms of its compressive strength. An addendum to SNiP 2.03.01-84 [11] covers a strength range from 2.5 to 40 N/mm$^2$ with densities from 800 to 2000 kg/m$^3$ depending on the type of aggregate used.

In the relevant European standard [12] lightweight aggregate concrete is classified according to density, as follows:

| Density class | 1.0 | 1.2 | 1.4 | 1.6 | 1.8 | 2.0 |
|---|---|---|---|---|---|---|
| Oven-dry density (kg/km$^3$) | 901–1000 | 1001–1200 | 1201–1400 | 1401–1600 | 1601–1800 | 1801–2000 |

## 2.2 Properties of lightweight aggregate for structural concrete

Such diverse materials as clay, PFA, shale, slag, etc., after processing using differing techniques produce aggregates with remarkably similar

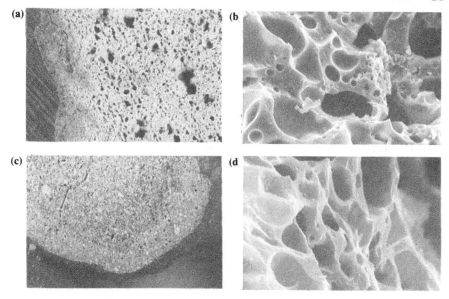

**Figure 2.1** Cross-sections and photomicrographs of typical lightweight aggregates. Expanded clay: (a) partial cross-section; (b) micrograph of interior. Sintered PFA: (c) partial cross-section; (d) micrograph of interior.

chemical compositions and structures [13]. Typically they have a ceramic-like dense matrix with included air voids and most have a relatively dense exterior shell with a more voided interior (see Figure 2.1).

In the UK all lightweight aggregates (LWAs) except foamed slag (BS 877: Part 2) [14], air-cooled blastfurnace slag (BS 1047) [15] and clinker (BS 1165) [16] are covered by BS 3797: Part 2 [17]. Sampling and testing methods are described in BS 3681: Part 2 [18]. Some typical properties of LWAs used for structural concrete are shown in Table 2.2. These should be compared with dense aggregates which have a density range 1200–

**Table 2.2** Typical properties of lightweight aggregates

| Type | Shape | Water absorption (% by mass) | Oven-dry loose bulk density* (kg/m$^3$) |
|---|---|---|---|
| Expanded clay | Rounded | 12–14 | 350–500 |
| Expanded slate | Angular/irregular | 10–15 | 560–720 |
| Expanded shale | Rounded | 12–14 | 500–800 |
| Pumice | Angular/irregular | 30–40 | 500–880 |
| Sintered PFA | Rounded | 9–15 | 800–850 |
| Pelletised expanded b.f. slag | Smooth/irregular | 3–5 | 850–950 |

* Coarse fraction of aggregate (fine fraction is denser).

$1900 \, kg/m^3$ (extra dense up to $2600 \, kg/m^3$) and a water absorption of approximately 0.5–2%.

Manufactured lightweight aggregates are usually free from deleterious chemicals and do not induce harmful reactions. However, they should be checked for carbon content (loss on ignition ≤4%) and sulphates ($SO_3$ ≤ 1%).

## 2.3 Properties of structural lightweight aggregate concrete

### 2.3.1 Fresh concrete

This topic is covered in detail in chapter 4 but a summary of the principal differences between structural lightweight aggregate concrete and normal weight concrete is given below.

The increased absorption, decreased density and range of available lightweight aggregates should be considered when designing a concrete mix. Of particular importance is water absorption. All aggregates, whether natural or artificial, absorb water at a rate which diminishes with time. Such absorption is important in that for unsaturated or partially saturated aggregate it will influence such properties of fresh concrete as workability (including pumpability) and density and also affect such hardened properties as density, thermal insulation, fire resistance and freeze/thaw resistance.

For an individual aggregate particle the amount of water absorbed and the rate of absorption depend primarily on (a) the pore volume, (b) the distribution of pores within the particle and (c) the structure of the pores (i.e. whether connected or disconnected). For lightweight aggregate particles, which have a relatively large pore volume, the rate of water absorption is likely to be much higher than for natural dense aggregates. However, the characteristics of the surface zone of aggregate particles have a large influence on absorption such that the disparity between natural and lightweight aggregates may not be as large as expected from the differences in density. For example, the sintered 'shell' around some particles (such as those of expanded or sintered aggregates) containing small, relatively disconnected pores, impedes the absorption process.

The water absorption of aggregates is usually expressed as the proportion of the oven-dry mass absorbed after 30 min and 24 h. For lightweight aggregates the 24 h value generally lies within the range 5–15% of the dry mass compared with about 0.5–2% for most natural aggregates. An approximate estimation of the correction to be made for water absorbed by lightweight aggregates during and immediately after mixing can be made on the basis of the 30 min absorption value which typically lies within the range 3–12% [19]. It should be noted that, as for natural aggregates, water absorptions for the finer grades of a material will generally be higher than those for the coarser.

Suitable mix proportioning procedures are described in detail elsewhere [19–22]. For a given grade of concrete the resulting mixes generally contain a higher cement content than for normal weight concrete, and the maximum attainable strength is governed by the type of aggregate used. It should be noted that 'workability' is underestimated by the slump test [23] and, as for normal weight concrete, cohesion is improved by the use of air entrainment. For pumped concrete workability should be assessed by the flow test [24].

In view of the effect of water absorption of lightweight aggregate, allowance should be made for a possible loss in workability between mixing and placing and during transportation. Admixtures are essential for pumping since water can be forced into aggregate particles by the pressure in the pipeline.

The processes of compaction and curing are no different from normal weight concrete but lightweight aggregate concrete is more tolerant of poor curing owing to the reserve of water held within the aggregate particles. When finishing the surface of lightweight aggregate concrete due consideration should be given to the possibility of flotation of the lightweight aggregate particles.

### 2.3.2    Density

The density of structural lightweight aggregate concretes can range from approximately 1200 to 2000 kg/m$^3$ compared with 2300 to 2500 kg/m$^3$ for normal weight concretes.

As the behaviour of lightweight aggregate concrete is closely related to its density, and as the density as well as the strength and durability are important to the designer, it is essential to define what is meant by the following terms [21].

*Fresh concrete density*    The bulk density of concrete when compacted to a practicable minimum air void volume.

*Oven-dry density*    The bulk density after drying for 24 h in air at 105°C.

*Air-dry density*    The density in equilibrium with a dry environment (moisture content of approx. 5–10% by vol.). Approximately equal to fresh density less 100–200 kg/m$^3$ for LW coarse and fine aggregates and fresh density less 50–100 kg/m$^3$ for LW coarse and dense fine aggregate.

*Saturated density*    Approximately equal to fresh density plus 100–120 kg/m$^3$.

It is suggested that the fresh concrete density is used as a basis for comparison.

The principal factors influencing density are [21]:

*Cement content* A 100 kg/m³ increase in cement gives approximately a 50 kg/m³ increase in density.

*Relative density of aggregates* A substitution of lightweight for dense fines increases the density by approximately 150–200 kg/m³.

*Entrained air* The resulting changes to the mix proportions decrease density by approximately 90 kg/m³.

*Moisture content of aggregates* A concrete made with water-saturated or partially saturated aggregates will have a higher fresh density.

*Environmental conditions* Density changes in response to wetting or drying.

### 2.3.3 Strength

*(a) Uniaxial compression.* As for normal weight concrete a wide range of aggregates produces a corresponding range of cube strengths. When comparing lightweight aggregate concrete with normal weight concrete it is important to consider the types of constituent materials in both cases (see Figure 2.2). Factors affecting strength include:

*Strength and stiffness of aggregate particles* Weaker particles require stronger mortars and thus higher cement contents. The 'ceiling' strength of concrete depends upon the type of aggregate. Excellent particle–matrix bond and similarity of particle and matrix moduli ensure that the matrix is used efficiently (see Figure 2.3).

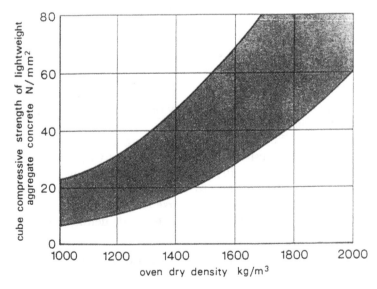

**Figure 2.2** Cube strength vs oven-dry density (ref. 79).

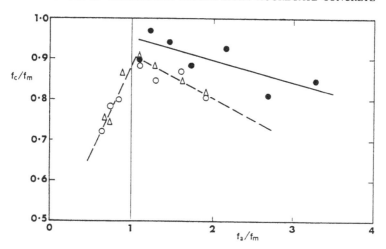

**Figure 2.3** Strength of concrete vs strength of aggregate (after Tanagawa *et al.*, 1977). $f_c$, $f_a$, $f_m$ = cylinder strengths of concrete, aggregate, paste; ● = river gravel aggregates; △ = lightweight aggregate (pelletised); ○ = lightweight aggregate (coated).

*Water/cement ratio* This has the same effect on strength as for normal weight concrete and the same range of water/cement ratio is used (see Figure 2.4). However, the reduction of effective water/cement ratio due to the water absorption of lightweight aggregate is difficult to predict and thus the specification of effective water/cement ratio for mixes is not practicable since it is difficult to measure and verify. Free water contents are the same as for normal weight concrete (say 180–200 l/m³) but aggregate absorption requires higher total water contents (say 250–300 l/m³). For lightweight aggregate concrete it is more relevant for mix design purposes to relate strength to cement content [19].

*Cement content* For a given workability, strength increases with cement content, the increase depending on the type of aggregates used (see Figure 2.5). Generally, greater cement contents are required than for normal weight concrete (approximately 10% greater below about 35–40 N/mm² and 20–25% above). Although the increase in strength for a given increase in cement content depends on the type of aggregate used and the cement content itself, on average for lightweight aggregate a 10% higher cement content will give approximately a 5% higher strength.

*Age* Similar strength–age relationships as for normal weight concrete (see Figure 2.6). If concrete dries then hydration will cease but the situation is better for lightweight aggregate concrete than for normal weight concrete owing to the reserve of water available in aggregate pores. Thus lightweight aggregate concrete is more tolerant of poor curing than normal weight concrete.

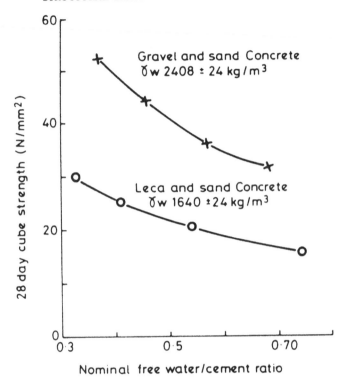

**Figure 2.4** Strength of concrete vs free water/cement ratio (ref. 21).

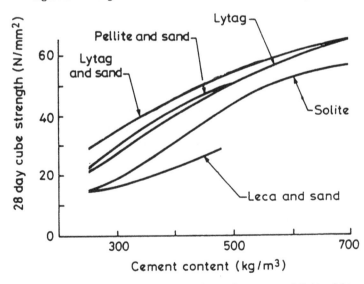

**Figure 2.5** Cube strength vs cement content for various types of lightweight aggregate concrete (ref. 21).

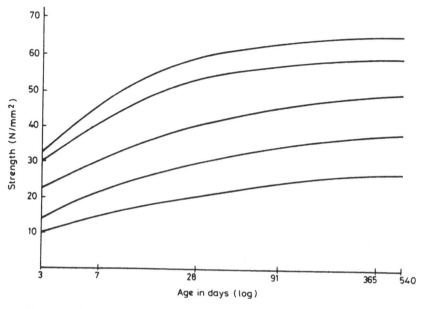

**Figure 2.6** Cube strength vs age for water-cured sintered PFA concrete (ref. 21).

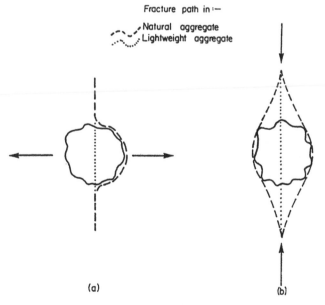

**Figure 2.7** Fracture paths in lightweight and normal weight concretes. ———— Natural aggregate; . . . . lightweight aggregate.

*Density*　The density of compacted concrete is affected mainly by aggregate density which is related to particle porosity and hence particle strength. Thus, aggregates of different density will result in different concrete strengths as well as densities (see Figure 2.4).

*(b) Tensile strength.*　Tensile strength is important when considering cracking in concrete elements. The factors influencing compressive strength also influence tensile strength. The principal differences between lightweight aggregate concrete and normal weight concrete are due to:

*Fracture path*　This travels through, rather than around, lightweight aggregate particles. The behaviour is similar to normal weight concrete made with crushed aggregates in that the flexural/compressive strength ratio is higher (see Figure 2.7).

*Total water content*　This is higher for lightweight aggregate concrete due to the absorption of the lightweight aggregate. Thus, in drying situations greater moisture gradients can cause a significant reduction in tensile strength although this effect is somewhat alleviated by the effects of increased hydration (see Figure 2.8). These effects should be considered when testing. The flexural strength is more affected than cylinder splitting strength.

For continuously moist cured concretes the measured tensile strength as a proportion of the compressive strength is similar for both lightweight and

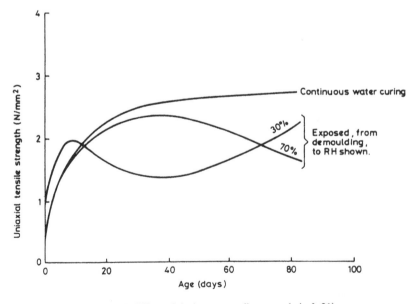

**Figure 2.8** Effect of drying on tensile strength (ref. 21).

normal weight concrete. However, for dried concrete the tensile strength of lightweight aggregate concrete is reduced below that for normal weight concrete of the same compressive strength. This reduction also influences other properties such as shear, bond, anchorage, etc.

BS 8110: Part 2 [65] requires that for Grade 25 concrete (i.e. concrete with not more than 5% of cube strength results less than 25 N/mm$^2$) and above, the design shear strength should not exceed 80% of the value for normal weight concrete (a table is given for Grade 20 concrete). In no case must the shear stress exceed the lesser of $0.63f_{cu}$ or 4 N/mm$^2$. However, a comprehensive investigation covering a wide range of lightweight aggregates has shown conclusively that this reduction is not justified for either shear [25] or punching shear [26].

*(c) Multiaxial compression.* All structural concretes show an increase in the axial stress required to cause failure if a lateral stress is applied simultaneously. Lightweight aggregate concrete exhibits the same trend but the strength enhancement is somewhat lower [27]. In view of this the following relationships have been proposed for lightweight aggregate concrete [28]:

$$\sigma_{1f} \leqslant 0.67f_{cu} + 2\sigma_3,$$

where $\sigma_1 \geqslant \sigma_2 \geqslant \sigma_3$ and,

$\sigma_{1f}$ = maximum principal compressive stress that can be sustained
$f_{cu}$ = characteristic cube strength
$\sigma_3$ = minor principal compressive stress.

The above relationship should be compared with that for normal weight concrete of:

$$\sigma_{1f} \leqslant 0.67f_{cu} + 3\sigma_3$$

Expressions are also given in the above reference for the limit states of serviceability and ultimate.

### 2.3.4 Strength/density ratio

By the very nature of structural lightweight aggregate concrete its strength /density ratio assumes a greater importance than for normal weight concrete. The following examples clearly demonstrate the advantages to be gained by using structural lightweight aggregate concrete for two types of structure.

1. A bridge in Germany constructed using both lightweight and normal weight concrete each containing approximately the same proportions of retarder and plasticising admixtures by mass of cement [29]:

|  | LWC | NWC |
|---|---|---|
| *Specification requirements* | | |
| Strength class | LB45 | B55 |
| Density (kg/m$^3$) | <1800 | 2400 |
| *Mix proportions* (kg/m$^3$) | | |
| Cement | 400 | 400 |
| Sand | 497 | 494 |
| Gravel | 83 | 706 |
| LWA | 623 | nil |
| Fly ash | 50 | 50 |
| Water | 170 | 172 |
| *Fresh state properties* | | |
| Density | 1880–1930 | 2380–2420 |
| Flow spread on site (mm) | 290–350 | 280–350 |
| *Cube strength* (N/mm$^2$ at 28 days) | | |
| Mean | 73.3 | 69.3 |
| Standard deviation | 4.7 | 7.5 |
| Characteristic | 65.6 | 57 |
| *Strength/density ratio* (N/mm$^2$/kg/m$^3$) × 1000 | | |
| Based on mean strength | 38 | 29 |
| Based on characteristic strength | 34 | 24 |

2. A comparative study of the use of lightweight aggregate concrete for offshore structures [30]:

|  | LWC | NWC |
|---|---|---|
| *Mix characteristics* | | |
| Cement content (kg/m$^3$) | 410 | 400 |
| Free W/C ratio | 0.40–0.42 | 0.40–0.42 |
| *Fresh state properties* | | |
| Fresh density (kg/m$^3$) | 1875 | 2420 |
| Slump (mm) | 150–200 | 150–230 |
| *Cube strength* (N/mm$^2$ at 28 days) | | |
| Range | 60–70 | 60–70 |
| Standard deviation | 4 | 4–5 |
| *Other hardened state properties* | | |
| Tensile splitting strength (N/mm$^2$) | 4.0 | 4.0 |
| Flexural strength (N/mm$^2$) | 6.0 | 6.0 |
| Ultimate creep at 100% RH (%) | 0.8 | 0.8 |
| Ultimate shrinkage at 65% RH (%) | 0.05 | 0.05 |
| Thermal expansion per °C | | |
| *Strength/density ratio* (N/mm$^2$ kg/m$^3$) × 1000 | | |
| Based on mean strength | 35 | 27 |

The above examples clearly demonstrate the improved strength/density ratios for the lightweight concrete despite the similar fresh and hardened properties for both types of concrete. Of particular interest is the improved control of the lightweight concrete as exemplified by the lower standard deviations. Such observations confirm the view that structural lightweight aggregate concrete should be treated no differently from normal weight concrete.

### 2.3.5 Impact

The lower modulus of elasticity and higher tensile strain capacity of lightweight aggregate concrete provides better impact resistance than normal weight concrete. This has been used to advantage in various situations, including the protection of underwater structures [31, 32] and stair treads [33]. More research is required if the full capabilities of the material are to be exploited.

### 2.3.6 Deformation

*(a) General.* Lower stiffnesses of lightweight aggregate particles and higher cement contents result in larger deformations. However, these effects are alleviated by lower density so small-scale tests in the laboratory can provide pessimistic data compared with behaviour on site.

The stress–strain relationships for lightweight aggregate concrete are more linear and 'brittle' than for normal weight concrete (see Figure 2.9). Such a response is probably attributable to the greater compatibility

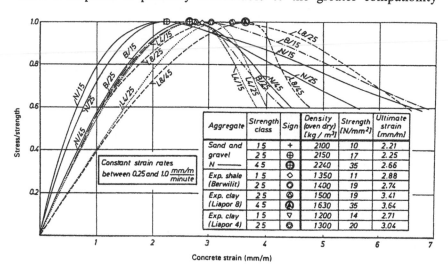

| Aggregate | Strength class | Sign | Density (oven dry) [kg / m³] | Strength [N/mm²] | Ultimate strain [mm/m] |
|---|---|---|---|---|---|
| Sand and gravel N | 1 5 | + | 2100 | 10 | 2.21 |
| | 2 5 | ⊕ | 2150 | 17 | 2.25 |
| | 4 5 | ⊕ | 2240 | 35 | 2.66 |
| Exp. shale (Berwilit) | 1 5 | ◇ | 1350 | 11 | 2.88 |
| | 2 5 | ◉ | 1400 | 19 | 2.74 |
| Exp. clay (Liapor 8) | 2 5 | ⊘ | 1500 | 19 | 3.41 |
| | 4 5 | ⊘ | 1630 | 35 | 3.64 |
| Exp. clay (Liapor 4) | 1 5 | ▽ | 1200 | 14 | 2.71 |
| | 2 5 | ⊖ | 1300 | 20 | 3.04 |

Constant strain rates between 0.25 and 1.0 $\frac{mm/m}{minute}$

**Figure 2.9** Stress–strain curves (ref. 79).

between the lightweight aggregate particles and the surrounding cementitious matrix. In the case of normal weight concrete the formation and propagation of small microscopic cracks, or microcracks (2–5 μm), have long been recognised as the causes of fracture and failure of concrete and the marked non-linearity of the stress–strain curve, particularly near the ultimate stress level. Although some of these discontinuities exist as a result of the compaction process of fresh concrete, the formation of small fissures or microcracks in concrete is due primarily to the strain and stress concentrations resulting from the incompatibility of the stiffnesses of the aggregate and cement paste/mortar components. The fracture process in normal weight concrete starts with stable fracture initiation (up to approximately 50% of ultimate stress) followed by stable and then unstable fracture propagation (from approximately 75 to 85% of ultimate stress) [34]. Although fundamentally the same process occurs in lightweight aggregate concrete the stable fracture initiation stage is extended and the unstable fracture propagation stage is reduced, so complete disruption occurs abruptly at ultimate. It should be noted that this behaviour is not accounted for in most codes of practice.

BS 8110: Part 2 [65] allows deflections to be calculated as for normal weight concrete with appropriate values for modulus of elasticity, free shrinkage and creep for lightweight concrete made with the given aggregate. Alternatively, heavily loaded members may be checked against the limiting span/effective depth ratios to 85% of those for normal weight concrete.

*(b) Modulus of elasticity, E-value.*   For any concrete its stiffness depends on the stiffnesses of the various constituents and their relative volumetric proportions in the mix. Simplifying concrete as a two-phase material consisting of coarse aggregate particles embedded in a mortar matrix, the *E*-value of the composite will decrease with (a) a decrease in the stiffness of the mortar which, in turn, depends on the volume proportions of cement, water and fine aggregate and the type of fine aggregate and (b) a decrease in the stiffness of the coarse aggregate. Since the moduli of lightweight aggregate particles are generally lower than those of natural dense aggregates and the fact that most lightweight aggregate concretes contain higher cement contents it follows that the overall moduli of lightweight aggregate concretes will be lower than normal weight concretes. It also follows that concretes made with lightweight coarse and lightweight fine aggregate will be lower than those made with lightweight coarse aggregate and natural dense fines.

It should be noted that although the *E*-value is not directly related to strength and density, a useful empirical relationship which provides an approximate *E*-value for most purposes is [21]:

$$E = D^2 \sqrt{f_{cu}} \times 10^{-6} \, kN/mm^2$$

where $E$, $D$ and $f_{cu}$ are modulus, nominal density and cube strength respectively. For calculating elastic deformation using BS 8110: Part 2 [65], the $E$-values are derived from the values for normal weight concrete ($E_{c,28} = K_0 + 0.2 \times f_{cu,28}$ where $K_0$ varies from 14 to $26 \, N/mm^2$) by multiplying by $(D/2400)^2$, where $D$ is the density of lightweight concrete.

Where more accurate values are required tests should be carried out on the given concrete.

The lower $E$-values for lightweight concretes will give rise to increased deformations for structural elements under a given load, although the effect will be reduced by the lower dead loads of lightweight concrete elements themselves. However, under dynamic conditions, such as impact or load fluctuation, reduced stiffness can be beneficial.

*(c) Creep.* Generally higher creep strains are produced in lightweight aggregate concrete than in normal weight concrete due to the lower $E$-value of aggregate and higher proportion of matrix.

The basic creep of lightweight aggregate concrete is approximately $1.00-1.15 \times$ that of normal weight concrete, but drying creep is often significantly higher. Thus, in full-scale structures, creep is seldom as large as predicted from laboratory tests due to differences in exposure conditions, curing, reinforcement, restraint, stress level, size, shape, etc.

*(d) Shrinkage.* As for creep, the laboratory data for shrinkage are pessimistic and do not relate to full-scale situations. Although shrinkage is generally greater for lightweight aggregate concrete made with lightweight fines than for normal weight concrete ($\times 1.0-1.5$) it is similar for lightweight aggregate concrete made with dense fines. Creep and shrinkage occur simultaneously and perhaps due to this effect shrinkage cracking is rare in lightweight aggregate concrete due to relief of restraint by creep, continuous supply of water from aggregate pores and better tensile strain capacity. This is recognised in BS 8110: Part 2, Table 3.2 [65]. Realistic values of drying shrinkage for full-scale structures subject to external exposure are 200–300 microstrain. For internal exposure these values may increase to 300–500 microstrain. A value of 350 microstrain could be assumed if no other information is available.

### 2.3.7   Bond and anchorage

The bond strength of lightweight aggregate concrete is similar to that of normal weight concrete but, as for normal weight concrete, excessive water contents should be avoided. Anchorage in lightweight aggregate concrete is lower than in normal weight concrete due to the lower bearing strength caused by the relative weakness of lightweight aggregates which requires careful detailing. BS 8110 [65] requires that the bond stresses used to determine lap and anchorage lengths for lightweight aggregate concrete be taken as 80% of those for the same grade of normal weight concrete. However, research shows that this factor is conservative [35].

For prestressed concrete transmission lengths are approximately 20–25% longer.

### 2.3.8   Fatigue

A comprehensive survey of test results under repeated compressive load showed that lightweight concrete above a density of $1500\,kg/m^3$ had the same susceptibility to fatigue as normal weight concrete when expressed as a Wohler or S–N diagram [36]. For both types of material the diagram could be described by the following equation:

$$f_{max}/f_{min} = 1 - 0.0685(1 - R)\,\log_{10} N$$

where, $f_{max}$ and $f_{min}$ are highest and lowest compressive (cylinder) stresses, $R$ is the ratio of lowest to highest compressive stress and $N$ is the number of load cycles to failure.

The above conclusion has been confirmed by other work [37] which showed lightweight aggregate concrete to perform at least as well as, and in some cases slightly better than, normal weight concrete of the same grade. However, other work has indicated that lightweight concrete made with sintered PFA aggregate performed less well than gravel concrete [38].

Information compiled by the CEB [39] shows that for repeated tensile stress states the fatigue strength of lightweight concrete is the same as or slightly higher than that of normal weight concrete [40, 41] while the fatigue strength under tensile–compressive states of stress is slightly lower [41].

As in the case of impact behaviour, more data are required on this subject.

### 2.3.9   Durability

(a) Freeze/thaw behaviour.   As for normal weight concrete the performance of lightweight aggregate concrete under freeze/thaw conditions depends mainly on the mix proportions, the type of aggregate and its moisture content and the level of air entrainment. Laboratory tests have shown that for the majority of aggregate types both in the presoaked and air-dry condition non-air-entrained lightweight aggregate concrete is potentially more durable under freeze/thaw conditions than equivalent strength non-air-entrained normal weight concrete [42, 43], particularly when natural fines are used [44]. For air-entrained concrete made with pre-soaked aggregates the performance of lightweight aggregate concrete is not significantly different from that of normal weight concrete, while air-entrained concrete made with lightweight aggregate in the air-dry condition shows a significant improvement over similar normal weight concrete [42]. High-strength lightweight aggregate concretes (54–73 N/mm$^2$) were found to exhibit 'outstanding performance' under standard freeze/thaw testing [45]. Prolonged exposure to simulated arctic offshore conditions was needed to cause significant damage, and behaviour was dependent mainly on moisture content and moisture condition of the aggregates.

*(b) Chemical resistance.*  Lightweight aggregates are stable since they are fired at approximately 1200°C and do not react with alkalis. However, most structural grades of lightweight aggregate concrete are made with dense fines and these should be checked for potential reactivity. In addition, the matrix has a lower free water/cement ratio and higher cement content which should reduce penetration.

*(c) Abrasion resistance.*  As with normal weight concrete, the resistance to abrasion of lightweight aggregate concrete increases with compressive strength [20]. However, if the matrix of lightweight aggregate concrete is abraded to expose the aggregate particles, it will deteriorate relatively rapidly. Resistance can be improved by combining relatively soft coarse aggregate with a hard fine aggregate, improving the quality of the matrix and the use of surface treatments.

*(d) Water absorption.*  Most lightweight aggregates exhibit significantly higher water absoptions than normal weight aggregates. This results in lightweight aggregate concretes having higher absorptions than typical normal weight concretes on a mass basis [20] although the difference is not as large as expected since the aggregate particles in lightweight aggregate concrete are surrounded by a high-quality matrix. It has been shown that water absorption of lightweight concrete does not relate directly to durability [46] and this may be due to the inappropriateness, when comparing concretes, of using absorption values based on mass rather than volume [21].

*(e) Permeability.*  Permeability or, more correctly, penetrability is the principal factor influencing durability. However, porosity and permeability are not synonymous since size of pores and their continuity must be taken into consideration.

Lightweight aggregate concrete is not necessarily more permeable than normal weight concrete [47] since porous lightweight aggregates are surrounded by a matrix which is less cracked [48] owing to (a) a lack of restraint from aggregate particles, (b) excellent aggregate–matrix bond [49, 50] due to the surface characteristics of the particles and pozzolanic action, (c) increased hydration of cement due to improved curing and (d) fewer heat of hydration effects [51]. It has also been suggested that improved permeability could be the result of a coating of dense cement paste around lightweight aggregate particles [52]. Researchers have confirmed this behaviour in measuring lower water permeabilities for air-cured lightweight aggregate concretes compared with normal weight concretes [62, 53] and much lower gas permeabilities [62, 54]. Tests on concretes cured under a variety of conditions from oven drying to sealed show structural lightweight aggregate concrete to have similar or lower oxygen permeability when compared with normal weight concrete of the same volumetric mix proportions [55]. This behaviour has been confirmed in additional tests [56], shown in Table 2.3.

Table 2.3

| Aggregate type | Resistivity (ohm-m) | Water permeability ($/10^{-12}\,m^2$) | Oxygen permeability ($/10^{-16}\,m^2$) |
|---|---|---|---|
| Leca | 650 | 5 | 0.5 |
| Lytag | 350 | 5 | 0.4 |
| Liapor | 600 | 15 | 0.4 |
| Granite | 500 | 85 | 1.0 |

Water and chloride penetration tests carried out on lightweight aggregate concretes with cube strengths from 88 to 104 N/mm$^2$ showed (a) penetrabilities which were low and independent of the porosity of the lightweight aggregate used [57] and (b) an optimum cement content above which penetrability was increased. However, some test results on lightweight aggregate concretes with strengths between 50 and 90 N/mm$^2$ show water and gas permeability to be slightly lower for lightweight aggregate concrete than for normal weight concrete [50].

*(f) Carbonation.* Carbonation is the reaction between carbon dioxide in the atmosphere, moisture and the minerals present in the cement paste. This reaction reduces the alkalinity of the concrete and can lead to shrinkage; but, more importantly, if it reaches any embedded metal such as reinforcement, it can promote the processes of corrosion.

Most lightweight aggregate particles are more porous and penetrable than normal weight aggregate particles and this, in common with normal weight aggregates of higher porosities [58], will allow more diffusion of gases such as carbon dioxide. However, if lightweight aggregate particles are well distributed in a good-quality matrix the rate of carbonation should be similar to that for normal weight concrete. Thus, it is essential to ensure that no continuous paths through particles exist between the surface and the reinforcement (the difference between the maximum and mean carbonation depth is approximately 0.5 × maximum size of aggregate) [59, 60]. For this reason, and from considerations of corrosion, BS 8110 [65] requires greater cover to lightweight aggregate concrete.

However, tests show the resistance to carbonation of lightweight aggregate concrete to be slightly better to slightly worse than for normal weight concrete of the same cement content or strength [43, 54]. Carbonation depth decreases with an increase in cement content and with the use of dense fines [59]. Exposure tests in a polluted atmosphere carried out on a range of concretes [61] indicated that for cement contents above $350 \, kg/m^3$ the carbonation depth was small for both lightweight aggregate concrete and normal weight concrete.

*(g) Corrosion of steel.* The higher cement content in lightweight aggregate concrete provides a highly alkaline environment to inhibit corrosion. This, together with the increased likelihood of achieving good compaction, reduces the corrosion risk [21]. However, low cement contents (less than approximately $300 \, kg/m^3$) may lead to early corrosion. Although BS 8110 [65] requires greater cover for lightweight aggregate concrete, performance in practice suggests that this may not be required in most cases.

### 2.3.10   *Thermal behaviour*

*(a) Thermal expansion.*   Coefficients of thermal expansion/contraction for lightweight aggregate concrete are less than for normal weight concrete made with the majority of aggregate types. As shown below, the range of coefficients for lightweight aggregate concrete is similar to that for normal weight concrete made with limestone aggregate [21].

| Type of concrete | Coefficient ($10^{-6}/°C$) |
|---|---|
| LWAC | −7 to −9 |
| NWC | −10 to −13 |
| NWC with limestone aggregate | −8 to −9 |

It has been suggested and confirmed by testing [62] that the combination of low thermal expansion and high tensile strain capacity results in a lower likelihood of lightweight aggregate concrete cracking under thermal stress, the critical temperature changes being approximately twice those for normal weight concrete. This phenomenon is taken into account by BS 8110 [65] where the temperature reduction allowed for lightweight concrete is 1.23 and 2.68 times the value for limestone and gravel concretes respectively.

*(b) Thermal conductivity.* Air in the cellular structure of structural lightweight aggregates reduces the rate of heat transfer compared with natural aggregates. Thus their inclusion within a cementitious matrix improves thermal conductivity. For this reason the good thermal properties of lightweight concretes have been used widely to improve thermal insulation in buildings and structures. The thermal conductivities of lightweight aggregate concretes vary primarily due to concrete density, aggregate type and moisture content [63]. Although this variation is known to exist, in the UK it is assumed that there are empirical relationships between conductivity and density for a given moisture content and class of material [64].

For concretes, whether lightweight or normal weight, conductivity values are given for densities from 400 to 2400 kg/m³ and moisture contents of 3% (protected environments) and 5% (exposed environments). These relationships are shown in Figure 2.10 and they appear to conform to best fit curves of the form $k = c^\lambda$, where $k$ is the thermal conductivity, $c$ is a

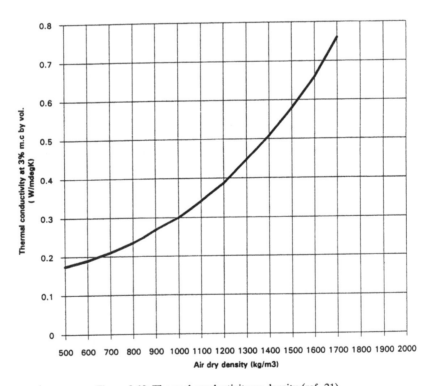

**Figure 2.10** Thermal conductivity vs density (ref. 21).

constant and $\lambda$ is the bulk dry density. The constants derived from the CIBSE data are as follows:

$$\text{Moisture content 3\%:} \ c = 0.085$$
$$\text{Moisture content 5\%:} \ c = 0.093$$

*(c) Temperature extremes.* Lightweight aggregate concretes exhibit excellent behaviour under fire and cryogenic conditions. This is significantly better than normal weight concrete and their high-temperature characteristics are acknowledged in BS 8110 [65] which requires lower covers to steel.

Structural lightweight aggregate concretes are suitable for high fire ratings and in this respect are superior to normal weight concretes since they (a) exhibit a lower strength reduction at high temperature, (b) provide more insulation and (c) have lower thermal expansions which result in less spalling.

Benefits can be achieved using lightweight aggregate concrete under cryogenic conditions such as for the storage of liquid gases [66, 67] due to its lower penetrability, higher strain capacity and, consequently, greater crack resistance and the enhancement of these properties at low temperatures.

### 2.3.11 Acoustic behaviour

The insulation of airborne sound transmission in solid homogeneous units such as walls and floors is improved as their mass increases. Thus, to achieve the same insulation the thickness of lightweight concrete units would need to be greater than normal weight concrete. However, tests have shown [68–70] that units made with lightweight concrete containing aggregates having a closely textured surface can perform better than predicted by their mass. For example, the sound insulation of a 200 mm thick wall made with expanded clay aggregate was found to be the same (52 dB) as that for a normal weight concrete wall of the same thickness. Also a 175 mm thick wall made with concrete (density 1600 kg/m$^3$) containing sintered pulverised-fuel ash gave the same insulation as a normal weight concrete wall (density 2400 kg/m$^3$) of the same thickness containing gravel aggregates.

### 2.3.12 Fire resistance

Structural lightweight aggregate concretes are more resistant to fire than normal weight concrete [20, 21] since (a) they experience a lower reduction in strength at high temperature due to the inherent stability of lightweight aggregate, (b) they provide more insulation due to their improved thermal conductivity (enabling cover to be reduced) and (c) they have reduced thermal expansions, causing less spalling.

## 2.4  Experience in use

Although the properties of lightweight aggregate concrete can be indicated by data obtained from laboratory testing, the performance of the material can only be demonstrated adequately by its performance in the field and, preferably, using full-scale elements or structures under service conditions.

A survey carried out on the use of lightweight aggregate concrete for marine applications [71] showed its successful use in shipbuilding since the First World War and its long-term durability for bridges, jetties and pontoons in the marine environment. A similar survey [72] also catalogued the successful use of lightweight aggregate concrete in a variety of fixed and floating structures and discussed its advantages and disadvantages. It was shown that lightweight aggregate concrete could withstand the rigorous conditions in marine and offshore situations and gave details of novel projects in which it could be used successfully.

Lightweight aggregate concrete has been used satisfactorily in bridges [73] and samples extracted from various bridge structures in North America showed the excellent performance of the material, due in part to the enhanced bond at the aggregate/paste interface and reduced microcracking as a result of the compatibility between aggregate and mortar [49]. It was concluded from this study that lightweight aggregate concrete had been able to resist satisfactorily several decades of severe exposure conditions. These views were confirmed by another report [74] which concluded that the use of lightweight aggregate concrete in highway structures was likely to result in improved durability and, consequently, reduced maintenance costs.

Long-term observations of a wide variety of lightweight aggregate concrete structures in Japan [75] showed that after 13–20 years of service no reduction was found in strength and modulus of elasticity and no increase in salt penetration and cracking. It was concluded that there was no inferiority compared with normal weight concrete.

Exposure tests carried out to compare lightweight aggregate concrete with normal weight concrete showed that reinforced lightweight aggregate concrete can be durable with regard to carbonation and corrosion for at least 28 years in an aggressive environment [76]. It was concluded that aggregate/cement ratio was more significant than the choice of aggregate and that lightweight aggregate concrete performed better than normal weight concrete at the same aggregate/cement ratio. Different conclusions may have been drawn had the comparisons been made on the basis of concrete grade and if the mixes tested had been made with modern materials.

Experimental work was carried out in the USA on air-entrained concrete slabs made with various types of normal weight aggregate and an expanded shale lightweight aggregate [77]. These were subjected to external exposure conditions including the repeated application of de-icing salt solution. After 202 freeze/thaw cycles and 123 applications of de-icer only 1 out of 17 lightweight aggregate concrete slabs was affected (with a slight blemish) while only 4 out of the 27 normal weight concrete slabs survived, these 4 having been surface treated.

A comprehensive examination of 40 structures in the UK [78] indicated that lightweight aggregate concrete is no less durable than normal weight concrete but might be more sensitive to poor workmanship. Problems identified in older structures, such as rate of carbonation and early age microcracking, would be overcome by modern practices of using cement contents of $400\,\text{kg/m}^3$ or more, natural fine aggregate and water reduction.

# References

1. Functional classification of lightweight concretes, *Recommendation LC2*, 2nd edition, RILEM, 1978.
2. Terminology and definitions of lightweight concrete, *Recommendation LC1*, 1st edition, RILEM, 1975.
3. *CEB-FIP Model Code for Concrete Structures*, Vol. II, 3rd edition, CEB, Paris, 1978.
4. Eurocode No. 2: Design of concrete structures, Part 1: General rules and rules for buildings, Final text, October 1991, EEC.
5. Eurocode No. 2 Part 1–4: The use of lightweight aggregate concrete with closed structure, Draft, June 1992, EEC.
6. AS 3600: Concrete structures, Standards Association of Australia, Sydney, 1988.
7. NS 3473: Concrete structures, Norwegian Council for Building Standardisation, 1989.
8. ACI 318–89: Requirements for reinforced concrete, ACI, Detroit, 1989.
9. Standard specification for design and construction of concrete structures, Part 1 (Design), Japan Society of Civil Engineers, Tokyo, 1986.
10. JASS 5 (revised 1979): Japanese architectural standard for reinforced concrete, Architectural Institute of Japan, March 1982.
11. SNiP 2.03.01-84: Manual for the design of concrete and reinforced concrete structures from normal weight concrete and lightweight concrete without prestressing the reinforcement, USSR.
12. ENV 206: Concrete – Performance, production, placing and compliance criteria, Final draft, February 1989. (BSI Document 89/11639)
13. Bremner, T. W. and Newman, J. B., Microstructure of low density concrete aggregate, *Proc. 9th Congress of FIP*, Vol. 3, Stockholm, June 1992, FIP, Slough, England.
14. BS 877: Specification for foamed or expanded blastfurnace slag lightweight aggregate for concrete, Part 2: 1973, BSI, London.
15. BS 1047: Specification for air-cooled blastfurnace slag aggregate for use in construction, 1983, BSI, London.
16. BS 1165: Specification for clinker and furnace bottom ash aggregates for concrete, 1985, BSI, London.
17. BS 3797: Specification for lightweight aggregates for concrete, Part 2:1976, BSI, London.
18. BS 3681: Methods for sampling and testing of lightweight aggregates for concrete, Part 2:1976, BSI, London.

19. Lydon, F. W., *Concrete Mix Design*, 2nd edition, Applied Science Publishers, 1982.
20. ACI Committee 213, Guide for structural lightweight aggregate concrete, *Journal ACI*, **64** (No. 8; 1967), 433–69.
21. I. Struct. E./Conc. Soc., *Guide to the Structural Use of Lightweight Aggregate Concrete*, I. Struct. E; London, October 1987.
22. ACI Committee 211-2-91: Recommended practice for selecting proportions for structural lightweight aggregate concrete, 1992.
23. BS 1881 (slump test): Method for determination of slump, Part 102: 1983, BSI, London.
24. BS 1881 (flow test): Method of determination of flow, Part 108: 1984, BSI, London.
25. Clarke, J. L., Shear strength of lightweight aggregate concrete beams, *Magazine of Concrete Research*, **39** (No. 141; 1987), 205–13.
26. Clarke, J. L. and Birjandi, F. K., Shear resistance of lightweight aggregate concrete slabs, *Magazine of Concrete Research*, **42** (No. 152; 1990), 171–6.
27. Hanson, J. A., *Strength of Structural Lightweight Aggregate Concrete under Combined Stresses*, Portland Cement Association, Development Bulletin D61, 1963.
28. Hobbs, D. W., Newman, J. B. and Pomeroy, C. D., Design stresses for concrete subjected to multiaxial stress, *The Structural Engineer*, **55** (No. 4; 1977), 151–64.
29. Grube and Knop, personal communication.
30. Norwegian contractors, personal communication.
31. Jensen, J. J., Impact of falling loads on submerged structures, *Proc of Int. Symp. on Offshore Structures*, Rio de Janeiro, Brazil, October 1979, pp. 1215–32.
32. Jensen, J. J. and Høiseth, K., Impact of dropped objects on lightweight concrete, *Nordic Concrete Research*, *Publication* No.2, Nordic Concrete Federation, Oslo, December 1983, pp. 102–13.
33. Bailey, J. H., Bentley S., Mayfield, B. and Pell, P. S., Impact testing of fibre-reinforced stair treads, *Magazine of Concrete Research*, **27** (No. 92; September 1975).
34. Newman, J. B., Concrete under complex states of stress, Chapter 5, *Developments in Concrete Technology – 1*. (ed. Lydon, F. D.), Applied Science Publishers, 1979.
35. Clarke, J. L. and Birjandi, F. K., *The Bond of Reinforcement in Lightweight Aggregate Concrete*, C&CA Services Report 1.039.00.2, BCA.
36. Tepfers, R. and Kutti, T., Fatigue strenth of plain, ordinary and lightweight concrete, *Journal of ACI. Proc.* **76** (No. 5; May 1979), 635–52.
37. Waagaard, K. and Kepp, B., Fatigue of high strength lightweight aggregate concrete, *Proc. of Symp. on Utilisation of High Strength Concrete*, Stavanger, Norway, June 1987, pp. 291–306.
38. Sparks, P. R., The influence of rate of loading and material variability on the fatigue characteristics of concrete, A.C.I. Special Publication SP-75, *Fatigue of Concrete Structures*, Detroit, 1982, pp. 331–41.
39. CEB, *Fatigue of Concrete Structures, State of the Art Report*, CEB Bulletin d'Information, No. 108, Lausanne, 1988, pp. 300.
40. Saito, M., Tensile fatigue strength of lightweight concrete, *Journal of Cement Composites and Lightweight Concrete*, **6** (No. 3; 1984), 143–9.
41. Cornellissen, H. A. W., *Fatigue of Concrete*, Part 4, *Tensile and Tensile -Compressive Stresses*, IRO-MATS/CUR Report, 1987.
42. Klieger, P. and Hanson, J. A., Freezing and thawing tests of lightweight aggregate concrete, *Journal of ACI*, **57** (No. 7; January 1961), 779–96.
43. Dhir, R. K., Durability potential of lightweight aggregate concrete *Concrete* (April 1980), 10.
44. Lydon, F. D., Some freeze-thaw test results from structural lightweight aggregate concrete, *Precast Concrete*, **12** (1981).
45. Whiting, D. and Burg, R., Freezing and thawing durability of high strength lightweight concretes, *SP 126-4*, pp. 83–100.
46. Hanson, J. A., Replacement of lightweight aggregate fines with natural sand in structural concrete, *Journal of ACI, Proc.* **61** (No. 7; July 1964), 779–94.
47. Soroka, I. and Jaegermann, C. H., Permeability of lightweight aggregate concrete, Paper No. 2.3.4, *2nd CEB/RILEM Symp. on Moisture Problems in Buildings*, Rotterdam, September 1974.
48. Hornain, H. and Regourd, M., Microcracking of concrete, *Proc. 8th Int. Congress on the Chemistry of Cement*, Rio de Janeiro, 1986, Vol. 5, pp. 53–9.

49. Holm, T. A., Bremner, T. W. and Newman, J. B., Lightweight aggregate concrete subject to severe weathering, *Concrete International - Design and Construction*, 6 (No. 6; June 1984), 49–54.
50. Bamforth, P. B., The properties of high strength lightweight concrete, *Concrete*, 21 (No. 4; April 1987), 8–9.
51. Lydon, F. D. and Al-Momen, M. H., The effects of moderate heat of hydration on some early and later age properties of concrete, *RILEM Int. Conf. on Concrete of Early Ages*, Paris, April 1982, p. 97.
52. Nishi *et al.*, Watertightness of concrete against sea water, *Journal of Central Res. Lab.*, Onoda Cement Co., Tokyo, 32 (No. 104; 1980), 40–53.
53. Dhir, R. K., Munday, J. G. L. and Cheng, H. T., Lightweight concrete durability, *Construction Weekly* (25 August 1989), 11–13.
54. Lydon, F. D. and Mahawish, A. H., Strength and permeability results for a range of concretes, *Cement and Concrete Research*, 19 (1989), 366–76.
55. Ben-Othman, B. and Buenfeld, N. R., Oxygen permeability of structural lightweight aggregate concrete, *Protection of Concrete* (eds Dhir, R. K. and Green, J. W.), E & FN Spon, 1990, pp. 725–36.
56. Denno, G., Imperial College (unpublished).
57. Zhang, Min-Hong and Gjorv, O. E., Permeability of high-strength lightweight concrete, *ACI Materials Journal*, 88 (No. 5; September/October 1991), 463–9.
58. Collins, R. J., Carbonation – comparison of results for concretes containing PFA, cementitious slag, or alternative aggregates, *Materials Science and Technology*, 3 (December 1987), 986–92.
59. Bandyopadhyay, A. K. and Swamy, R. N., Durability of steel embedded in structural lightweight concrete, *RILEM, Materiaux et Constructions*, 8 (No. 45, 1975).
60. Schulze, W. and Gunzler, J., Corrosion protection of the reinforcement in lightweight concrete, *Proc. 1st Int. Congress on Lightweight Concrete*, London, May 1968, C&CA, 1970.
61. Grimer, F. J., Durability of steel embedded in lightweight concrete, *Concrete*, 1 (No. 4; April 1967), 125–31.
62. Bamforth, P. B., *The Performance of High Strength Lightweight Concrete using Lytag Aggregate*, Taywood Engineering Research Report No. 014H/85/282, July 1985.
63. FIP State of Art Report, *Principles of Thermal Insulation with Respect to Lightweight Concrete*, FIP, Report FIP/8/1, C&CA, Slough, England, 1978.
64. CIBSE Guide A3, *Thermal Properties of Building Materials*, CIBSE, London, 1986.
65. BS 8110: Structural use of concrete: Part 2: Code of practice for special circumstances, British Standards Institution, London, 1985.
66. Bamforth, P. B., Murray, W. T. and Browne, R. D., The application of concrete property data at cryogenic temperature to LNG tank design, *2nd Int. Conf. on Cryogenic Concretes*, Amsterdam, October 1983 (organised jointly by UK and Dutch Concrete Societies).
67. Berner, D., Gerwick, B. C. and Polivka, M., Prestressed lightweight concrete in the transport of cryogenic liquids, *Oceans '83 Conference*, MTS and IEEE, San Francisco, August/September 1983.
68. Schule, W., Functional properties of concrete and concrete structures, *Zement-Taschenbuch, 1968/69*, Verein Deutscher Zementwerke EV, Wiesbaden, Bauverlag GmbH, 1968, pp. 331–56.
69. Veronnaud, L., Physical properties of lightweight concrete, *Revue des Materiaux de Construction et des Travaux Publics* (No. 62; November 1970), 338–42.
70. Forder, C., Lightweight concrete's place in the insulation spectrum, *Concrete* (January 1975).
71. FIP, State of Art Report, *Lightweight Aggregate Concrete for Marine Structures*, FIP, 1982.
72. Concrete Society, Technical Report No. 16, *Structural Lightweight Aggregate Concrete for Marine and Offshore Applications*, The Concrete Society, May 1978, 29pp.
73. Raithby, K. and Lydon, F. D., Lightweight concrete in highway bridges, *Int. Journal of Cement Composites and Lightweight Concrete*, 2 (No. 3; May 1981), 133–46.
74. Concrete Society, *Design and Cost Studies of Lightweight Concrete Highway Bridges*, Report of an investigation by the Lightweight Concrete Committee, April 1985.

75. Ohuchi, T. *et al.*, Some long term observation results of artificial lightweight aggregate concrete for structural use in Japan, *RILEM/ACI Int. Symp. on Long Term Observation of Concrete Structures*, *Budapest*, September 1984, pp. 273–82.
76. Nicholls, J. C. and Longland, J. T., *The Durability of Reinforced Lightweight Aggregate Concrete*, Dept of Environment, Building Research Establishment, Nore No. 80/87, August 1987, 32pp.
77. Walsh, R. V., Restoring salt-damaged highway bridges, *ASCE, Civil Engineering*, **37** (No. 5; May 1967), 57–9.
78. Mays, G. C. and Barnes, R. A., The performance of lightweight aggregate concrete structures in service, *The Structural Engineer*, **69** (No. 20; October 1991), 351–61.
79. *FIP Manual of Lightweight Aggregate Concrete*, 2nd ed., Surrey University Press, 1983, 259pp.

# 3 Design requirements

## J. L. CLARKE

## 3.1 Provision for lightweight aggregate concrete in codes

### 3.1.1 Introduction

This chapter reviews the design of lightweight aggregate concrete in various structural codes and identifies areas in which special provisions are made. The code requirements are, where appropriate, compared with test data and suggestions made for more realistic values. In general, no attempt is made to compare the actual design values given by the various codes but only the design approaches, to show how lightweight aggregate concrete is treated differently from normal weight concrete. Direct comparison between codes is difficult because they include different safety factors, which are not always clearly expressed in the design equations. Where sufficient test evidence is available, comparisons have been made with the design approaches in British codes.

Some codes take the concrete strength as that measured on cubes while others use cylinders. For the purposes of this chapter, cube strengths have been used throughout in line with current British practice. Where necessary, cylinder strengths have been converted to cube strengths by dividing by 0.85. Some structural aspects are covered in more detail in subsequent chapters.

The use of lightweight aggregate concrete is included in most structural design codes. Some limitations are applied and the information is either presented in the form of a separate section or else added into the relevant design clauses. The latter is a more reasonable approach: lightweight aggregate concrete should be seen as simply a variation on normal density concrete, although some of its properties will be different.

The strengths and densities of lightweight aggregate concrete vary considerably. Figure 3.1 shows data from tests carried out at the British Cement Association [1] on concretes using a range of European aggregates. It may be seen that strengths as high as $78 \, \text{N/mm}^2$ were obtained and air dry densities as low as $1450 \, \text{kg/m}^3$. It is interesting to compare the resulting strength/weight ratios with that of a 'normal' concrete, having a strength of $30 \, \text{N/mm}^2$ and a density of $2500 \, \text{kg/m}^3$. Typical values are shown in Table 3.1. This immediately suggests that any limitations on the

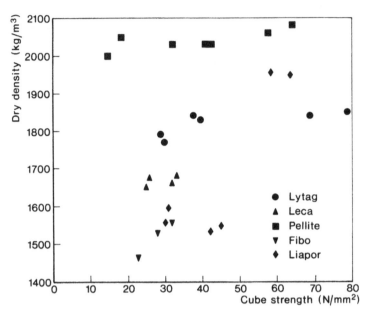

**Figure 3.1** Density and strength for various lightweight aggregates.

**Table 3.1** Comparison of strength and densities from work at BCA [1]

| Aggregate | Strength (N/mm²) | Density (kg/m³) | Strength/weight ratio* |
|---|---|---|---|
| Lytag | 70 | 1850 | 3.2 |
| Pellite | 55 | 2050 | 2.2 |
| Liapor | 45 | 1550 | 2.4 |
| Leca | 30 | 1650 | 1.5 |
| Fibo | 25 | 1500 | 1.4 |

*Compared with 30 N/mm² normal weight concrete with a density of 2500 kg/m³.

load-carrying capacity of lightweight aggregate concrete that may be imposed by design codes will be offset, to a greater or lesser extent, by the reduced self-weight of the structure.

### 3.1.2 British codes

BS 8110 [2] permits the use of lightweight aggregate concrete. It says in the section giving additional considerations in the use of lightweight aggregate concrete:

**Figure 3.2** Footbridge on the Kenilworth Bypass.

In considering lightweight aggregate concrete, the properties of any particular type of aggregate can be established far more accurately than for most naturally occurring materials. The engineer should therefore obtain specific data direct from the aggregate producer in preference to using tabulated values taken from British Standard codes of practice or specifications.

This is a rational approach. Because of the limited range of lightweight aggregate available in a particular area, the designer will be working with a particular material in mind, with its attendant properties. The same will not necessarily hold for a designer working with normal weight aggregate, particularly in major city areas. For example, in London at present readymixed concrete suppliers are using a range of aggregates, including local Thames Valley gravel, limestone from the Mendip Hills and granite from Scotland. At the design stage, the engineer may well not know which supplier will be used, and hence which aggregate, and so will be forced to use standard tabulated values.

The same statement is included in the code of practice for the design of bridges, BS 5400 [3]. Figure 3.2 shows an elegant lightweight concrete footbridge.

Two British codes of practice prohibit the use of lightweight aggregate concrete, namely BS 8007, 'Design of concrete structures for retaining aqueous liquids' [4] and BS 6349, Part 1, 'Maritime structures' [5]. Both codes work on the erroneous assumption that the porosity of the lightweight aggregate will lead to a permeable concrete.

### 3.1.3  American codes

In contrast, the American Concrete Institute 'State-of-the-art report on offshore concrete structures for the Arctic' [6] permits the use of lightweight aggregate concrete. In fact, it states that: 'Lightweight aggregate concrete has a proven record in the marine environment.' Similarly, the 'State-of-the-art report on barge-like concrete structures' [7] makes extensive reference to the use of lightweight aggregate concrete. It points out that high-strength, low-weight and high-durability lightweight concretes are available. A strength of $70 \, \text{N/mm}^2$ with a density of about $1800 \, \text{kg/m}^3$ can be achieved. The report also notes that ships, barges and jetties constructed using lower strength lightweight concretes during the Second World War are still in service, showing little deterioration.

The design clauses in both reports refer to the American Concrete Institute building code (see below). The use of lightweight aggregate in floating and marine structures is discussed in chapter 7.

The American Concrete Institute building code, ACI 318, and its commentary [8], require the use of the tensile strength of the concrete in some clauses, though a tabulated value may be used in the absence of the actual value. Interestingly, ACI 318 defines two types of concrete, namely 'lightweight concrete' in which lightweight material is used for both the fine and coarse aggregate and 'sand–lightweight concrete' in which natural sand is used for all the fine aggregate. Properties of the latter are close to those of normal aggregate. Partial replacement of the fines is also permitted, with consequent interpolation between the relevant provisions.

### 3.1.4  Norwegian code

The recently revised Norwegian code for concrete structures, NS 3473 [9], applies to all concrete, with both normal and lightweight aggregates, provided they fall within specified strength grades. The maximum grades, $105 \, \text{N/mm}^2$ for normal weight concrete and $85 \, \text{N/mm}^2$ for lightweight concrete, are considerably higher than those given, or implied, by most other design codes. Any type of aggregate, or combination of lightweight and natural aggregate, may be used, provided the resulting concrete has an oven-dry density in excess of $1200 \, \text{kg/m}^3$. However, the code adds a requirement that the various properties of any lightweight aggregate, such as strength, density, degree of burning, etc., should be uniform and that the bulk density should not vary by more than 7.5% from the specified value.

### 3.1.5  European code

Design of structures throughout Europe will eventually be covered by a European code. The main section, Part 1, which has been issued in draft form as ENV 1992–1–1 [10] gives general rules and rules for buildings, which specifically exclude lightweight aggregate concrete. Part 1C, 'The use of lightweight aggregate concrete with closed structures', is currently being drafted. This states that all clauses of Part 1 apply unless they are substituted by special clauses in Part 1C. It applies to all natural or artificial lightweight aggregates, unless reliable evidence indicates that provisions different from those given can be adopted safely.

### 3.1.6 Australian code

The Australian code for concrete structures [11] is applicable also to lightweight aggregate concrete having a surface dry density of not less than $1800 \, kg/m^3$. The commentary to the code, Supplement 1, says that most Australian concretes employ only lightweight coarse aggregate resulting in concrete with a surface dry density that is seldom less than $1800 \, kg/m^3$. Concrete made from both coarse and fine lightweight aggregates are specifically excluded from the code. In the absence of specific rules, designers apparently turn to the ACI code for guidance.

Interestingly, apart from the section outlining the scope of the document, the code makes no reference to the type of aggregate being used. Hence, it does not differentiate between lightweight aggregate concrete and normal weight concrete, provided the density is above $1800 \, kg/m^3$.

### 3.1.7 Japanese specifications

The Japanese Society of Civil Engineers specification [12] makes fairly extensive reference to lightweight concrete. This is taken to mean concrete with both coarse and fine lightweight aggregates, i.e. using the same terminology as the American codes. Though the specification makes reference to combinations of lightweight and natural aggregates, and states that the properties will depend on the ratio of types of aggregate used, values are only given for all lightweight aggregate concrete.

The Architectural Institute of Japan specification [13] gives five types of lightweight concrete, depending on the strength and type of aggregate and on its intended use. The specification covers the production of the concrete and certain aspects of construction but does not deal with the design of structures.

## 3.2 Design requirements for reinforced concrete

### 3.2.1 Introduction

The sequence of presentation of the design clauses in various codes is somewhat different. The order used in BS 8110 has been adopted for the comparisons in this section.

### 3.2.2 Definition of lightweight concrete

Lightweight aggregate concrete in British codes [2, 3] is taken as having an oven-dry density of less than $200 \, kg/m^3$. Any concrete with a density greater than that is treated as normal weight. In the draft European code [10] density classes from 1.0 to 2.0 are given for lightweight aggregate concrete. Typically, density class 1.6 refers to an oven-dry density of $1401-1600 \, kg/m^3$ with corresponding design densities of $1650 \, kg/m^3$ for plain concrete and $1750 \, kg/m^3$ for reinforced concrete. The ACI definition of structural lightweight concrete [8] is one having an air-dry density of less than $1810 \, kg/m^3$. This would exclude some stronger, but heavier, European aggregates, which would thus class as normal weight concretes for design purposes.

The Australian code [11] covers structural lightweight aggregate concrete with a minimum density of $1800 \, kg/m^3$. Those with lower densities are specifically excluded.

The Norwegian code [9] defines lightweight aggregate concrete as having an oven-dry density between 1200 and $2200 \, kg/m^3$. For design purposes, the average density for reinforced concrete is taken as the oven-dry density plus $150 \, kg/m^3$.

The Japanese specifications [12, 13] do not define densities for lightweight concrete.

### 3.2.3 Limitations on compressive strength

All structural codes of practice define the minimum strength of concrete that may be used, generally from considerations of durability. For structural lightweight aggregate the strengths given in Table 3.2 apply, all values being in $N/mm^2$. The Japanese JASS 5 [13] defines high-grade and common-grade structures. It has been assumed that the former is more appropriate for this comparison.

**Table 3.2** Minimum strengths for structural lightweight concrete

| Code | | Reinforced | Prestressed |
|---|---|---|---|
| BS 8110 | [2] | 20 | 30 [post-tensioned] <br> 40 [pretensioned] |
| BS 5400 | [3] | 25 | Not permitted |
| ACI 318 | [8] | 30 | 30 |
| ENV 1992–1–4 | [10] | 12 | 25 [post-tensioned] <br> 30 [pretensioned] |
| AS 3600 | [11] | 25 | 25 |
| NS 3473 | [9] | 25 | 35 |
| JASS 5 | [13] | 25 | 25 |

Maximum strengths are not specified in British codes, though they are generally limited to $50\,N/mm^2$ for reinforced concrete and $60\,N/mm^2$ for prestressed concrete. No specific limitations are set for lightweight aggregate concrete. The European code [10], on the other hand, specifically limits the strength to $60\,N/mm^2$ for both reinforced and prestressed concrete. The Norwegian code [9] sets an upper limit of $85\,N/mm^2$ for lightweight aggregate concrete. In addition, the strength may not exceed $105 \times (w/200)^{1.5}$, where $w$ is the oven-dry density. Thus a $1500\,kg/m^3$ concrete would be limited to $59\,N/mm^2$ while an $1800\,kg/m^3$ concrete could be up to $78\,N/mm^2$. The Australian code [11] does not appear to specify a maximum strength that may be used though the tabulated values extend only to $60\,N/mm^2$. JASS 5 [13] sets no upper limit but the Japanese Society of Civil Engineers specification [12] gives a maximum strength in the tables of $47\,N/mm^2$.

### 3.2.4   Cover to reinforcement

*(a) Code requirements.*   BS 8110 [2] defines five exposure conditions, ranging from mild to extreme. The former corresponds to concrete protected against the weather and the latter to surfaces exposed to abrasive action. In all cases, apart from mild, the code requires an additional 10 mm of cover when lightweight aggregate concrete is used, compared with that required for an equivalent grade of normal weight concrete. The same requirement is given in BS 5400 [3]. The additional cover is based on the assumption that, because the lightweight aggregate is more porous than normal weight natural aggregates, then the resulting concrete must similarly be more porous. However, this is not substantiated by test data (see below), nor is there evidence of lightweight aggregate concrete structures being less durable than normal density concrete ones.

ACI 318 [8], AS 3600 [11] and NS 3473 [9] do not differentiate between different concretes when determining the cover requirements, neither does the draft European code 2, Part 1C [10], though it does emphasise that the cover in lightweight aggregate concrete is more sensitive to poor workmanship than is normal weight concrete and, hence, special care is required.

The two Japanese specifications are somewhat contradictory. The Japanese Society of Civil Engineers [12] makes no reference to the type of concrete. However, JASS 5 [13] requires 10 mm more cover for lightweight concrete in contact with soil; no adjustment is required for concrete not in contact with soil.

*(b) Research evidence.*   Mays and Barnes [14] reviewed data from research on concretes using aggregates available in the UK. They reported

studies on the permeability of water, nitrogen and oxygen. In all cases, the permeability was found to be less than that of normal weight concrete with the same strength, sometimes by two or three orders of magnitude. As far as carbonation is concerned, they concluded that there was little difference.

The most significant part of their study was an inspection of 40 structures in the UK, all constructed before 1977 and all having exposed structural lightweight concrete elements. The structures fell into five groups – namely, bridges, car parks, *in-situ* buildings, precast buildings and 'other' structures. The latter category included Mulberry Harbour units dating from 1944, grandstands and storage structures. Figures 3.3 and 3.4 show two of the structures considered. A visual inspection of the 40 structures identified four that were showing significant signs of reinforcement corrosion. This was generally due to low cover, possibly combined with poor detailing. Others showed less severe signs of deterioration, again generally linked to poor detailing or to thermal effects, but 40% of the structures were virtually defect-free. Six of the structures were inspected in more detail, including measurement of average carbonation depth, corrosion potential, *in-situ* strength by means of cores, permeability, etc.

In the light of their observations, and limited data from the USA, Canada, Japan and Australia, the authors concluded that:

> There is no evidence from the results of this investigation that lightweight aggregate concrete is any less durable than normal weight concrete.

Thus it would appear that the additional cover requirements in British codes are unnecessarily conservative.

**Figure 3.3** Student residences, University of East Anglia.

**Figure 3.4** Car park and bus station, Rochdale.

### 3.2.5 Fire

*(a) Code requirements.* Lightweight aggregate concrete gives a better performance in fire than normal weight concrete. More detailed information may be obtained from the Institution of Structural Engineers/Concrete Society guide [15]. Simply put, the improved behaviour is due to three factors:

1. During the manufacturing process lightweight aggregates are subjected to high temperatures, in excess of those experienced in a conventional fire, and hence are more stable than many naturally occurring aggregates.
2. Lightweight aggregate concrete has a lower coefficient of expansion and a lower elastic modulus than normal weight concrete. In addition, it has a higher tensile strain capacity, all of which reduces the tendency for the formation of microcracks.
3. The reduction in strength at elevated temperature is significantly less than in normal weight concrete. At 800°C it retains about 40% of its ambient strength while normal weight concrete would be only about 20%.

While these three factors lead to improved performance of the exposed concrete, the lower thermal conductivity and diffusivity leads to a slower rise in temperature in the body of the member and hence a reduced effect on the reinforcement.

The improved performance of lightweight aggregate concrete in fire is reflected in the relevant clauses of BS 8110 [2], which specifies reduced minimum dimensions and reduced cover requirements, particularly for higher durations. In addition, the requirements for anti-spalling steel for situations in which high covers are essential are less onerous for lightweight.

The Australian code [11] makes no specific reference to lightweight aggregate concrete in its section on design for fire resistance. However, fire resistance may be determined by calculation, using appropriate values for Australian materials which should show an advantage for lightweight aggregate. The fire-resistance period for slabs may be increased by adding a lightweight aggregate concrete topping.

The Norwegian code [9] again makes no specific reference to lightweight concrete in the sections dealing with fire. It gives general information on the reduction in strength of concrete at elevated temperatures, which may be used for short-duration fires. If the concrete is exposed to temperatures above 200°C for a long period of time, then actual properties of the concrete must be used. (The code does not define short- and long-term exposure.) Thus, once again, lightweight aggregate concrete should show an advantage.

The Japanese Society of Civil Engineers specification [12] makes only a brief mention of cover requirements for fire protection. It says that the cover will depend on the characteristics of the aggregate used but gives no further guidance.

*(b) Behaviour of structures in fire.* There is only limited published information on the behaviour of lightweight aggregate concrete buildings in fire. Forrest [16] attempted to obtain information from 25 countries but found that generally 'no national authority distinguishes between dense and lightweight aggregates when reporting on fires in concrete structures'. He does publish details of fires in six lightweight buildings (in London, Melbourne and Japan) and concludes that the 'fire resistance has been extremely good and the building restored to usefulness with a minimum of repair or even no repair necessary'.

A widespread use of lightweight aggregate concrete is in composite floors, with profiled metal decking. Lawson [17] reported the results of fire tests on full-scale slabs, using both normal weight concrete and lightweight concrete with Lytag (sintered PFA) aggregate. He reported that the lightweight concrete slab showed no significant spalling, while the normal weight slab spalled severely in the early stages of the fire. This lack of spalling is confirmed by Malhotra [18] who concluded that 'concretes made with lightweight aggregates are shown to be virtually free from spalling in fire tests'. However, he noted that spalling can occur if the moisture content is high.

### 3.2.6 Flexure

The British and American codes do not differentiate between lightweight and normal weight concrete when considering the strength of members in flexure. However, the draft European code [10] gives special provisions for lightweight aggregate concrete. It defines an idealised bilinear stress–strain diagram for concrete, with a peak stress of 0.77 times the design strength of the concrete in most situations. For normal weight concrete the factor is 0.85. The transition from the linearly increasing portion of the curve to the uniform is at a strain of 0.00135 for normal weight concrete, but increases to 0.0022 for lightweight aggregate concrete. The maximum strain is 0.0035 for all concretes. Thus in flexure, the stress block is closer to triangular in shape, if these idealised diagrams are used. As an alternative, a simple rectangular stress block may be used, with the same peak stress, the depth being 0.8 times the neutral axis depth as for normal weight concrete. For under-reinforced sections these changes have no significant effect on the ultimate moment capacity. However, for over-reinforced sections the moment capacity, for a given grade of concrete, will be slightly reduced.

The Norwegian code [9] also uses a bilinear stress–strain diagram for lightweight aggregate concrete, but only for grades between 15 and 45 N/mm$^2$. The transition is at 0.002. For higher grades a parabolic–rectangular diagram is required, with the initial stiffness and the transition point determined by testing.

The Japanese Society of Civil Engineers specification [12] uses a parabolic–rectangular diagram for both normal and lightweight concrete. However, it does point out that a more appropriate curve should be used when determining deformations and ductility. It says that a more triangular diagram should be used with a sudden reduction in stress beyond the peak. No further specific guidance is given.

### 3.2.7 Shear resistance of beams

(a) Code requirements. In line with many codes, BS 8110 [2] assumes that the total shear resistance of a beam is the sum of the shear capacity of the concrete cross-section and that of the shear reinforcement. The shear capacity of the concrete cross-section when using lightweight aggregate is taken as 0.8 of the value for the equivalent grade of normal weight concrete. A similar reduction factor is applied by BS 5400 for bridges [3]. Both codes assume that the behaviour of the shear reinforcement is unaffected by the type of aggregate and both require minimum shear reinforcement equivalent to 0.4 N/mm$^2$ on the cross-section.

The maximum shear stress that may be applied to the cross-section is

0.8 times that for normal weight concrete. This is only likely to be a limitation for slender members.

The ACI building code [8] determines the shear capacity of the concrete cross-section on the basis of the tensile strength of the concrete, taken as equal to the square root of the compressive strength, $\sqrt{f'_c}$. For lightweight aggregate concrete the actual tensile strength, $f_{ct}$, is used, if known, replacing $\sqrt{f'_c}$ by $f_{ct}/6.7$, provided this latter term is smaller. When the tensile strength is not known, $\sqrt{f'_c}$ is multiplied by 0.75 for 'all lightweight' concrete and 0.85 for 'sand–lightweight' concrete. Linear interpolation is permitted when partial sand replacement is used. These factors are similar to those in the British codes. The code requires minimum shear reinforcement equivalent to $0.34\,\text{N/mm}^2$ ($50\,\text{lb/in}^2$), which is slightly less than in British codes.

The same multiplying factors are applied to the maximum shear that may be carried by the cross-section.

The Norwegian code [9] calculates the shear capacity on the basis of the tensile strength of the concrete. Values for the *in-situ* tensile strength are tabulated for the various grades of normal concrete. For lightweight aggregate concrete they are multiplied by

$$(0.3 + 0.7w/2400),$$

where $w$ is the oven-dry density. The resulting reduction factors are listed in Table 3.3.

For members without shear reinforcement, the shear capacity is given by

$$0.33(f_{td} + r)bd,$$

where $r$ is the percentage of properly anchored tension reinforcement, divided by the partial safety factor of 1.4. Thus, the net reduction factor for the shear capacity will depend on both the density of the concrete and the amount of tensile reinforcement. For example, for a $45\,\text{N/mm}^2$ concrete with a density of $1800\,\text{kg/m}^3$ and 2% of reinforcement, the reduction factor would be 0.9 not the 0.83 suggested by Table 3.3.

**Table 3.3** Tensile strength reduction factors in Norwegian code [9]

| Density (kg/m³) | Tensile strength reduction factor |
|---|---|
| 2200 | 0.94 |
| 2000 | 0.88 |
| 1800 | 0.83 |
| 1600 | 0.77 |
| 1400 | 0.71 |
| 1200 | 0.65 |

**Table 3.4** Tensile strength reduction factors in the draft European code [10]

| Density class | Maximum oven-dry density (kg/m³) | Factor |
|---|---|---|
| 2.0 | 2000 | 0.9 |
| 1.8 | 1800 | 0.85 |
| 1.6 | 1600 | 0.8 |
| 1.4 | 1400 | 0.75 |

The minimum shear reinforcement is a function of the tensile strength of the concrete, with a minimum value of about $0.5\,N/mm^2$. The calculation of the shear capacity of the concrete cross-section in the draft European code [10] is also based on the tensile capacity of the concrete. In the absence of more accurate data, the code requires that the specified tensile strength for a given grade of concrete should be modified by a factor

$$(0.4 + 0.6c/2400),$$

where $c$ is the density class, giving the factors listed in Table 3.4.

The maximum shear stress for lightweight aggregate concrete is about 85% of that for normal weight concrete.

The requirement for minimum shear reinforcement is based on the concrete strength and on the grade of steel used. Typically, for a $30\,N/mm^2$ concrete $0.47\,N/mm^2$ would be required, slightly higher than BS 8110.

The draft European code gives two methods for the design of members with shear reinforcement. The standard method assumes that the stirrups form part of a 45° internal truss, and the total capacity is equal to that of the concrete plus that of the stirrups. The alternative approach, the variable strut inclination method, is a more exact representation of the mechanism of shear failure. The resulting equations give the total capacity directly and depend only on the compressive strength of the concrete. Thus using this approach there would be no difference between lightweight and normal weight concrete.

In the Japanese Society of Civil Engineers specification [12] the shear capacity of the concrete section alone is based on the cube test of the concrete strength, as in British codes. For lightweight concrete the shear stress is multiplied by 0.7. The specification does not define a minimum amount of shear reinforcement.

The Australian code [11] makes no reference to lightweight aggregate concrete in the section dealing with shear and hence, provided the density is greater than $1800\,kg/m^3$, no adjustment is required.

The reduction factors applied to the tensile strength of equivalent normal weight concrete in the various codes are compared in Figure 3.5.

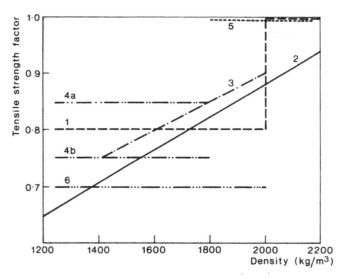

**Figure 3.5** Comparison of tensile strength reduction factors. Key: 1 BS 8110 [2]; 2 NS 3473 [9]; 3 ENV 1992–1–4 [10]; 4 ACI 318 [8]; 5 AS 3600 [11]; 6 JSCE [12].

*(b) Effect of minimum shear reinforcement.* In practice the provision of minimum shear reinforcement means that the shear capacity for lightweight aggregate beams is closer to that of normal weight beams than, for example, the 0.8 factor in BS 8110 [2] would suggest. This is illustrated by the following example:

Consider a 400 mm deep beam, with 1% main reinforcement, 30 N/mm² concrete and minimum shear reinforcement.

|                              | Lightweight | Normal |
|------------------------------|-------------|--------|
| Design shear stress          | 0.54        | 0.67   |
| Minimum shear reinforcement  | 0.40        | 0.40   |
| Total                        | 0.94        | 1.07   |

Hence reduction factor is 0.88 instead of 0.8.

With higher amounts of shear reinforcement the reduction will be less.

In the design process, the effects of the reduced shear capacity will be offset by the lower self-weight. In many circumstances the latter will be more significant than the former. Comparisons have been carried out, again for 1% of main reinforcement and 30 N/mm² concrete assuming that the lightweight has a density equal to 75% of the normal weight concrete. Figure 3.6 shows the ratios of the spans for lightweight aggregate and normal aggregate beams, based on shear capacity, for various amounts of shear reinforcement and various levels of live loading. It may

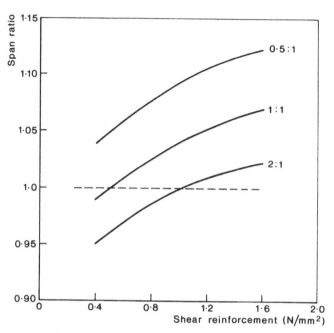

**Figure 3.6** Ratio of lightweight aggregate span to normal weight aggregate span for different live load/self-weight ratios. *Note*: Numbers on curves refer to live load/self-weight (normal concrete) ratio.

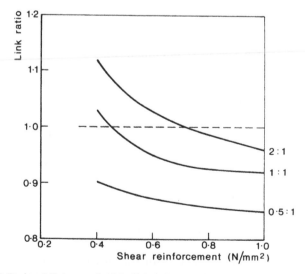

**Figure 3.7** Ratio of links required in lightweight beam to those in normal weight beam. Note: Numbers on curves refer to live load/self-weight (normal concrete) ratio.

be seen that in most cases a longer span is possible with the lightweight aggregate concrete.

Alternatively, the amount of shear reinforcement required for a given span is compared in Figure 3.7. For most of the cases considered the lightweight aggregate beam requires less than its normal weight counterpart.

Although the comparisons have been made on the basis of BS 8110, similar conclusions would be reached using any of the codes discussed above.

*(c) Research evidence.*   Much of the data on which the codes were based was derived from tests carried out in the USA in the 1960s [19] with limited work being done in the UK at the same time. Since that time manufacturing processes have been improved, leading to higher strength aggregates. Thus a reappraisal of the rules may be appropriate. Clarke [20] reviewed available data and carried out additional tests on beams, both with and without shear reinforcement. Two artificial lightweight aggregates were used as well as a pumice. The resulting densities were between 1700 and 2000 kg/m$^3$ and the strengths between 30 and 55 N/mm$^2$. He concluded that without shear reinforcement a 0.9 reduction factor would be more appropriate. For beams with links, no reduction factor was necessary. This is consistent with the comparisons made above. Replacing 0.8 by 0.9 leads to an effective reduction factor, with nominal shear reinforcement, of 0.94, which is of little significance.

### 3.2.8   Torsion

Design for torsion in British codes [2, 3] is carried out when the torsional shear stress exceeds a specified minimum which is a function of the concrete strength. Above this level torsional reinforcement and additional longitudinal reinforcement is required, which carries all the torque. The torsional shear stress, when added to the normal shear stress, may not exceed a maximum value which is again a function of the concrete strength. When lightweight aggregate concrete is used, a factor of 0.8 is again applied to both the minimum and maximum torsional stresses. Thus torsional reinforcement will be required at a lower load than with normal weight concrete. However, in general the amount of reinforcement will not be affected by the type of concrete.

The Japanese specification [12] would appear to use the same approach as the British codes, but a factor of 0.7 is applied when using lightweight concrete.

A similar approach is used by the Norwegian code [9], which limits the torsional shear stress to a function of the tensile strength of the concrete. Thus, in line with the values given in Table 3.3, lightweight concrete will require torsional reinforcement at a lower level of stress than normal weight concrete. Above this level, all the torque is carried by torsional reinforcement and additional longitudinal reinforcement, uninfluenced by the type of concrete being used.

ACI 318 [8], on the other hand, uses an approach similar to that for shear, the total torsional capacity being that of the concrete cross-section plus that of the reinforcement. The same reduction factors apply as for shear.

Again, the Australian code [11] requires no adjustment for lightweight aggregate concrete, provided the density is above $1800\,\mathrm{kg/m^3}$.

The draft European code [10] design approach for torsion is similar to the variable strut inclination method for shear and again depends only on the compressive strength of the concrete. Thus the type of aggregate will have no influence on the torsional capacity.

### 3.2.9 Deflections

BS 8110 [2] states that the deflection of beams may be checked by direct calculation. In this case the appropriate values for the modulus of elasticity, the free shrinkage and the creep coefficient may be used. In the absence of actual data, the modulus of elasticity may be obtained by multiplying the appropriate figure for normal weight concrete by $(w/2400)^2$, where $w$ is the density of the lightweight concrete in $\mathrm{kg/m^3}$. The draft European code [10] uses the same expression. BS 5400 [3] uses a similar approach but with 2300 as the denominator.

The Australian code [11], the Norwegian code [9] and the American code [8] all take the modulus of elasticity as being proportional to $[\mathrm{density}]^{1.5}$, again taking the density of normal weight concrete as $2400\,\mathrm{kg/m^3}$. Thus these codes would suggest that lightweight aggregate concretes are stiffer than the British codes assume (see Figure 3.8).

In the Japanese Society of Civil Engineers specification [12] the modulus of elasticity of lightweight aggregate concrete is taken as about 60% of that for the equivalent grade of normal weight concrete, irrespective of the density.

As an alternative to direct calculation, BS 8110 suggests that the span/effective depth ratios given for normal weight concrete elements may be used, except that where the characteristic imposed load exceeds $4\,\mathrm{kN/mm^2}$ the limiting span/effective depth ratio should be multiplied by 0.85.

Similarly, ACI 318 [8] gives minimum overall depths as a function of the span for various situations. For lightweight concrete with a unit weight $w$ in the range $1420-1890\,\mathrm{kg/m^3}$ ($90-120\,\mathrm{lb/ft^3}$) the depth must be multiplied by $(1.65 - 0.005w)$. The resulting factors are shown in Table 3.5.

The draft European code [10] also considers lightly stressed and highly

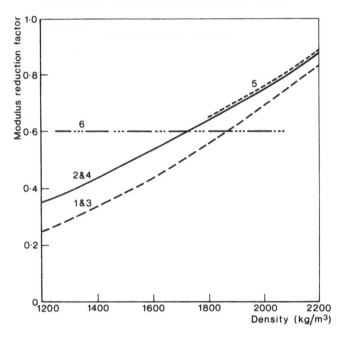

**Figure 3.8** Comparison of reduction factors for elastic modulus. For key see Figure 3.4.

**Table 3.5** Span/effective depth reduction factors in ACI 318 [8]

| Density (kg/m³) | Factor 1.65 − 0.005w | Span/effective depth reduction factor |
|---|---|---|
| 1800 | 1.08 | 0.93 |
| 1600 | 1.14 | 0.88 |
| 1400 | 1.21 | 0.83 |

stressed members, based on the level of stress in the reinforcement. As a guide it says that members with less than 0.5% of tension reinforcement can be taken as being lightly stressed, as can slabs normally. Highly stressed members are taken as having 1.5% of reinforcement or more. For intermediate amounts of reinforcement, intermediate factors may be taken. The span/effective depth ratios for lightweight aggregate concrete are 0.85 times those for normal weight concrete in all cases.

Where deflections are not critical, the Norwegian code [9] makes no distinction between normal and lightweight concrete.

Thus the British, American and European codes give very similar values for the reduction factor.

The Australian code [11] gives an expression for determining span/depth ratios, which depends on the modulus of elasticity of the concrete and the effective design loading, all to the power one-third. The resulting span/depth ratio reduction factor will depend on the level of loading and on the density of the concrete but will be similar to those in the other codes.

### 3.2.10  Shear of slabs

The British codes, the Norwegian code, ACI 318 and the draft European code all apply the same modification factor as for beams when considering the shear of slabs, either along a line across the slab or round a concentrated load. Thus, for slabs without shear reinforcement, they suggest that a lightweight slab will have approximately 80% of the shear capacity of a normal weight slab of equivalent concrete strength. The Japanese specifications [12, 13], which are for lightweight coarse and fine aggregate concrete, apply a factor of 0.7 to the normal weight strength.

Clarke and Birjandi [1] tested 26 lightweight aggregate slabs without shear reinforcement, using a total of five different aggregates and with strengths ranging between about 20 and 65 N/mm$^2$. They showed that the ratios of the actual punching shear strength to the value predicted by BS 8110 were similar in all cases. As the air-cured densities varied between about 1450 and 2050 kg/m$^3$, this suggests that the linkage between tensile strength and density given in the European code is also unnecessary. The lightweight aggregate specimens gave punching strengths similar to those of equivalent grade normal weight concrete slabs. From the results the authors concluded that the 0.8 reduction factor in BS 8110 is unnecessary and should be eliminated.

### 3.2.11  Columns

BS 8110 [2] gives very minor modifications to the design clauses for columns to take into account the effect of the lower modulus of elasticity on the determination of the slenderness. Short columns are taken as those with an effective height to cross-sectional dimension of less than 10, while for normal weight concrete the figure is 15. Similarly, the additional moment induced in the column by its deflection is about 65% higher for a lightweight aggregate column than in its normal weight counterpart.

These simple modifications would appear to have a major shortcoming. The modulus of elasticity of lightweight concretes varies significantly, depending on the density and the type of aggregate. However, the same corrections are made for all lightweight aggregate concretes. Even knowing the actual modulus, the modification factors cannot be adjusted.

The clauses in BS 5400 [3] make no specific reference to lightweight aggregate concrete. However, when determining the forces and moments in slender columns, the additional moments induced by deflections must be taken into account. Here the reduced stiffness of lightweight aggregate concrete may be of significance.

The ACI code [8] does not specifically mention lightweight aggregate concrete in its column design clauses. However, the moment magnification approach, which amplifies the applied column moments to take account of the effect of axial load on these moments, does take account of the modulus of elasticity of the concrete. Similarly, the Japanese Society of Civil Engineers specification [12] does not refer to the type of concrete when assessing column slenderness.

The design of columns in the draft European code [10] includes the modulus of elasticity of the concrete in the expression for determining the effective length of the column. However, it would appear to be significant only when the concrete in the columns differs from that in the remainder of the frame. Further information is included in Appendix 3 of the code, in which the modulus of elasticity of the concrete is considered as well as the creep. Both factors will affect the slenderness limits. Similarly, the Norwegian code [9] includes the modulus of elasticity in the expression for determining the additional eccentricity and also in the frame analysis.

It would appear that only limited experimental work has been carried out on lightweight aggregate concrete columns. However, available information agrees with the codes that, for short columns at least, there is little difference from the behaviour of normal weight concrete columns. Hoff *et al.* [22] tested a series of $50 \, \text{N/mm}^2$ lightweight aggregate concrete columns and compared the results with those from normal weight columns. They concluded that slightly higher margins of safety were achieved with the lightweight aggregate concrete, probably because the lower stiffness permitted complete mobilisation of the potential stress in the reinforcing steel.

### 3.2.12  Walls

For walls, BS 8110 [2] gives similar modifications to those applied to columns, again on account of the reduced elastic modulus. In BS 5400 [3], slender walls are designed in the same way as columns, as discussed above.

### 3.2.13  Detailing of reinforcement

*(a) Code requirements.*  The bond stresses used in BS 8110 [2] to determine the lap and anchorage lengths are taken as 80% of those for an equivalent grade of normal weight concrete. The Institution of Structural Engineers/Concrete Society guide [15] suggests that the restriction is based on inadequate evidence.

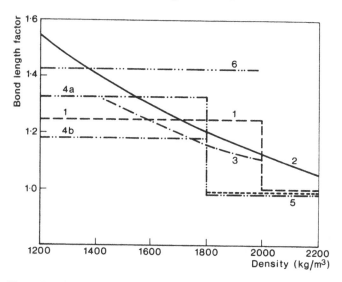

**Figure 3.9** Comparison of bond length factors. For key see Figure 3.4.

In the American code [8] the lap length and anchorage length are again related to the tensile strength of the concrete. Lengths may be based on actual tensile strengths, if known, but not less than the lengths for an equivalent grade of normal density concrete, or else increased by 1/0.75 for all lightweight mixes or 1/0.85 for sand–lightweight mixes as before. Similarly, in the draft European code [10] the tensile strength factor (see section 3.2.7 and Table 3.4) is used to modify the bond stresses. In addition, the minimum radii of bends are increased by 30%, to reduce bearing stresses. The Japanese Society of Civil Engineers specification [12] applies a reduction factor of 0.7 to bond stress, in the absence of more precise information.

In the Norwegian code [9], the design bond stress for ribbed bars is a function of the tensile strength of the concrete and of the amount of transverse reinforcement. The tensile strength is a function of the density (see Table 3.3).

Figure 3.9 compares the basic bond lengths required by the various codes. For a given density, the bond length for normal weight concrete is multiplied by the relevant factor to give the bond length for lightweight aggregate concrete.

*(b) Test evidence.* One difficulty is that there is no internationally agreed method for determining the bond strength of reinforcement. The various 'standard' methods that are used are generally compliance tests to check the adequacy of the reinforcement, which is cast into concrete that is

confined by secondary reinforcement. The tests do not represent the behaviour in a typical situation. Using a test apparatus developed at the British Cement Association, which more accurately represents the case of reinforcing bars in the corners of beams, Clarke and Birjandi [21] tested high-yield reinforcement in four different types of lightweight aggregate concrete. They compared the results with those from normal weight specimens and concluded that the 0.8 factor was unnecessarily conservative.

### 3.3  Design requirements for prestressed concrete

#### 3.3.1  Introduction

The introductory clause at the start of section 4 in BS 8110 [2], dealing with design and detailing of prestressed concrete, permits the use of lightweight aggregate concrete and refers to section 5 of Part 2, which gives general guidance on its use. However, no specific guidance is given apart from a general warning that prestress losses will be greater than those for dense aggregate concrete. The code of practice for the design of bridges, BS 5400 [3] prohibits the use of lightweight aggregate concrete for prestressed structures. This is a severe, and surprising, limitation bearing in mind the bridges and other major prestressed structures that have been built around the world (see chapters 6 and 7).

The clauses for prestressed structures in ACI 318 [8] refer back mainly to the equivalent clauses for reinforced concrete. No specific mention is made of lightweight concrete in the prestressed clauses and hence it may be assumed that the only adjustments required are those concerned with the tensile strength of the concrete, as discussed before.

The draft European code [10] makes little reference to prestressed concrete, which is generally considered as an additional loading on reinforced concrete. (A further part of the European code specifically dealing with prestressed concrete is being prepared.) Thus, as with the other codes, the design changes relate to the tensile strength of the concrete.

The following sections show how the clauses that specifically deal with prestressed concrete in the various codes may be adopted for use with lightweight aggregate concrete and, hence, how the resulting values are effected.

#### 3.3.2  Cover to reinforcement for durability and fire

The cover requirements in BS 8110 [2] for prestressed concrete to meet ductility requirements are the same as for reinforced concrete and hence, presumably, the same additional 10 mm is required when using lightweight aggregate. In ACI 318 [8] the durability requirements are as for normal weight concrete.

For fire protection to the reinforcement, BS 8110 again gives reduced minimum member sizes and reduced covers when using lightweight aggregate.

### 3.3.3 Service and transfer conditions

The permissible stresses in tension are based on the tensile strength of the concrete, expressed as a function of the square root of the compressive strength in both BS 8110 [2] and ACI 318 [8]. For lightweight aggregate concrete the expressions should be multiplied by 0.8 in BS 8110 and replaced by the tensile strength of the concrete in ACI 318, as discussed in section 3.2.7.

### 3.3.4 Shear of beams

As with reinforced members, the shear capacity of prestressed lightweight aggregate beams may be a cause for concern. BS 8110 [2] identifies two types of potential failure. The shear capacity of sections uncracked in flexure, given in the equation below, is based on the principal tensile stress being equal to the tensile capacity, $f_t$, of the concrete.

$$V_{co} = 0.67bh\sqrt{(f_t^2 + 0.8f_{cp}f_t)},$$

where $f_{cp}$ is the compressive stress due to prestress at the centroidal axis. For dense concrete $f_t$ is taken as $0.24\sqrt{f_{cu}}$. Applying a reduction factor 0.8 to the tensile strength gives an amended value of $0.2\sqrt{f_{cu}}$ for $f_t$. The reduction on the shear capacity of the concrete section will depend on the level of stress due to prestress at the centroid. In addition, minimum shear reinforcement equivalent to $0.4\,\text{N/mm}^2$ is required. Hence the net reduction in sections with minimum reinforcement will be significantly less than the 0.8 factor. For example:

Consider a beam with $50\,\text{N/mm}^2$ concrete and stress due to prestress of $8\,\text{N/mm}^2$.

| | Lightweight | Normal |
|---|---|---|
| Design shear stress | 2.17 | 2.50 |
| Minimum shear reinforcement | 0.40 | 0.40 |
| Total | 2.57 | 2.90 |

Hence reduction factor is 0.89 instead of 0.8.

This is similar to the net reduction factor for reinforced beams.

The design of sections cracked in flexure is based on an empirical equation derived from tests on normal weight members:

$$V_{cr} = [1 - (0.55f_{pe}/f_{pu})]v_c b_v d + M_0 V/M$$

where $f_{pe}/f_{pu}$ is the effective stress in the tendons divided by the characteristic strength and $v_c$ is the shear stress for an equivalent reinforced concrete section. $M_0$ is the moment necessary to produce zero stress at the extreme tension fibre and $V$ and $M$ are the applied shear force and moment at the section under consideration. However, because of the lack of experimental evidence, the handbook to BS 8110 [23] recommends that both terms in the expression be multiplied by 0.8. This is a conservative approach. The first term might reasonably be multiplied by 0.8 when using lightweight aggregate. However, no adjustment would be required to the second term, which simply depends on the level of stress at the extreme fibre due to the prestress and is presumably not influenced by the type of concrete. Reynolds *et al.* [24] shows that the second term is considerably larger than the first, typically perhaps seven times as large. This would suggest that there would be no significant difference between normal weight and lightweight beams of the same dimensions and with the same level of applied stress.

In the draft European code [10] the shear capacity of prestressed beams is treated as a special case of the shear of reinforced beams. The total shear capacity is a function of the tensile capacity of the concrete (see section 3.2.7 and Table 3.4), and the average level of stress due to prestress. Thus, for reasonable levels of prestress the difference between the shear capacity of lightweight aggregate concrete and normal weight concrete beams will be small.

### 3.3.5  Prestress losses

BS 8110 identifies a number of factors that cause a reduction in the prestressing force effective at any stage, as follows:

(a) relaxation of the tendon
(b) elastic, creep and shrinkage deformations in the concrete
(c) losses at the anchorage
(d) friction along the tendon.

In general, changing from normal concrete to lightweight aggregate concrete will only affect item (b). The only exception would appear to be when the tendon passes through an unlined duct. Here, because the lightweight aggregate pellets are generally weaker than natural aggregates, it is likely that the coefficient of friction will be slightly reduced.

As discussed in section 3.2.9, BS 8110 gives guidance on the elastic modulus to be used in the absence of actual test data, the values for normal weight concrete being multiplied by $[w/2400]^2$, where $w$ is the density in kg/m$^3$. The Australian code uses the same expression but with the power 2 replaced by 1.5. BS 8110 gives no specific guidance on suitable values for the creep and shrinkage but Spratt [25] suggests that the creep of structural lightweight aggregate concrete is only slightly greater than that of normal weight concrete. The inclusion of natural sand in the mix will reduce the creep. For general design purposes he suggests that the creep should be taken as between 1.3 and 1.6 times that of normal weight concrete. This is in line with the Japanese Society of Civil Engineers specification [12] which gives creep factors for lightweight concrete at different ages and for different environmental conditions. These are about 30% higher than the equivalent normal weight factors.

The Norwegian code [9] says that the creep factor may be assumed equal to that of normal weight concrete multiplied by

$$(w/2400)^{1.5} \quad \text{for } w \text{ greater than } 1800 \, \text{kg/m}^3$$
$$1.2(w/2400)^{1.5} \quad \text{for } w \text{ less than } 1500 \, \text{kg/m}^3,$$

where $w$ is the oven-dry density. For intermediate values, linear interpolation may be used. The resulting factors reduce with reducing density, and are very much lower than the values suggested by Spratt. (Possibly this is a typographical error in the translation.)

Similarly, Spratt mentions that the shrinkage of lightweight aggregate concrete will be higher, but again it is reduced by the use of natural sand. However, for general design work he suggests a value between 1.4 and 2 times that of normal weight concrete. In contrast the Japanese Society of Civil Engineers specification says that the same values may be used for both normal weight and lightweight concrete.

The figures given in the various codes are generally in line with the brief statement in the handbook to BS 8110 [23], which warns that overall loss of prestress can increase by as much as 50% when using lightweight aggregate.

### 3.3.6 *Transmission length*

No guidance is given in BS 8110 on the transmission length to be used with pretensioned members. However, the BS 8110 handbook [23] recommends that it be increased by 50% in the absence of appropriate test data.

## 3.4 Thermal effects

### 3.4.1 *Early thermal cracking during construction*

Practical aspects of construction using lightweight aggregate are covered in a later chapter. However, some relevant information is given in section

2 of Part 2 of BS 8110 [2] and also in BS 8007 [4] regarding early thermal cracking. The codes identify two situations in which thermal stresses occur, caused by movements due to the heat of hydration being resisted by restraint. The first is due to the thermal gradients that may occur in a thick member. Here the low thermal conductivity of the concrete prevents the heat in the centre of the member escaping to the surrounding atmosphere. The resulting expansion of the core may lead to tensile stresses at the surface, which may lead to cracking. Conversely, when the core cools the opposite effect may occur, causing internal cracking of the core.

The second is caused by external restraint – for example, when a wall is cast on a rigid base or a slab is cast against an existing slab. The heat of hydration will again cause the concrete to expand. However, at this stage it has little stiffness and any stresses caused by restraint will be rapidly relieved by creep. Having reached maximum temperature the element will cool and contract. It now has a significant stiffness and tensile stresses will be induced. The magnitude of the stresses will depend on the temperature difference between peak and ambient, the coefficient of expansion of the concrete and the degree of restraint.

The concrete will crack when the stresses exceed the tensile capacity of the concrete. BS 8110 gives limiting values of tensile strain capacity, thus avoiding the problem of assuming the elastic modulus at early ages. Comparing lightweight aggregate made of PFA (Lytag or similar) with gravel aggregate, the code shows an increase in tensile capacity of about 60% and also a decrease in coefficient of expansion of about 40%. This results in limiting temperature differentials for lightweight aggregate concrete that are about 2½ times those for a gravel concrete under the same conditions of restraint and the same temperature differential.

Harrison [26] gives lower values for the coefficient of expansion of lightweight aggregates than the code and also higher tensile strain capacities. This would lead to limiting temperature differentials about five times those for gravel concrete. However, Harrison notes that for a given specified strength, lightweight aggregate concrete will have a somewhat higher cement content than the equivalent normal weight concrete and, hence, will experience a higher temperature rise.

### 3.4.2 Thermal movements in mature concrete

Though not generally considered in the design of buildings, temperature changes on the mature concrete may induce significant stresses in the finished structure. These temperature changes, which will be most important for exposed buildings such as multistorey car parks, are due to daily and seasonal variations.

Guidance for bridges is given in Part 2 of BS 5400 [3], which specifies seasonal variations and differential temperatures between the top and bottom surface. The temperatures depend on the type of bridge and its location. As an example, a concrete bridge located in central England would experience a seasonal temperature change from −11°C to +36°C, and a daily differential, between the top surface and the interior, of about 11°C. On a 25 m span, the seasonal change would lead to a free movement of about 14 mm if a gravel aggregate is used but only 8 mm if a lightweight aggregate is used. The differential temperature will set up internal stresses, depending on the level of restraint. With its lower coefficient of expansion and its lower modulus of elasticity, lightweight aggregate concrete structures will experience significantly lower induced moments.

The design temperature ranges and differentials should be equally applicable to multistorey car parks and similar buildings. However, here there is likely to be less provision for movements than in a bridge and hence induced stresses will be much higher. Evans and Clarke [27] monitored the thermal movements in a car park over a period of 18 months. Cracks had formed at the tops of many of the columns just below the top deck. The authors recorded temperatures that were very similar to the predictions of BS 5400 and concluded that the cracks in the columns had been caused by thermal movements. Once the cracks had formed, the tops of the columns in effect became hinges, with appreciably less restraint than before. Rotations at these positions during the monitoring period agreed well with the observed temperatures. Once again the reduced coefficient of expansion of lightweight aggregate concrete, coupled with its higher tensile strain capacity and lower modulus of elasticity, would have significantly reduced the risk of cracking due to temperature variations.

## 3.5 Overall design implications

### 3.5.1 Introduction

The preceding sections, which reviewed various structural design codes, suggest that lightweight aggregate with a dry density of, say, 1800 kg/m$^3$ or greater should be treated as normal weight concrete. The various reduction factors – for example, those applied to the shear capacity of the concrete – will have little effect at these densities. For lighter concretes it will be necessary to apply the reduction factors. However, in many cases, the reduction factors will be offset by the reduced self-weight of the structure. This will be particularly true for the lower density concretes now available. The design of various types of structure is considered in subsequent chapters. However, various general conclusions may be drawn when comparing lightweight aggregate structures with those of equivalent normal weight. The most significant are listed below.

### 3.5.2 Cover to reinforcement

British codes, unlike other design codes, require higher covers to ensure durability. This is not justified by evidence from the behaviour of actual structures. The behaviour of lightweight aggregates in fire is superior to that of normal weight aggregates resulting in reduced covers.

### 3.5.3 Flexure

There will be no difference in the bending moment capacity for situations in which the steel is yielding. For compression failures in the concrete the moment capacity may be reduced slightly.

### 3.5.4 Shear of beams

Most codes apply a reduction factor to the concrete shear capacity though test evidence suggests this is probably unnecessary. Lightweight aggregate beams with shear reinforcement, i.e. all but beams of very minor importance, will carry at least 90% of the capacity of equivalent normal weight beams.

### 3.5.5 Shear of slabs

From the codes, the shear capacity of lightweight aggregate slabs without shear reinforcement will be about 80% of that of equivalent normal weight slabs. Limited research evidence suggests that this factor is conservative.

### 3.5.6 Deflections

For lightly loaded beams and slabs the maximum span/effective depth ratios will be unaltered. For more heavily loaded members they will be reduced by 15%. In practice it is found that actual deflections are less than the predicted values.

### 3.5.7 Columns

The reduced stiffness will lead to some additional moments in columns. The transition from short to slender column will be at a lower value of effective height to cross-sectional dimension. This is an area that requires experimental study.

### 3.5.8 Detailing

Anchorage and lap lengths of reinforcing bars in lightweight aggregate concrete will be about 25% longer than in normal weight concrete. This is probably conservative. The radius of bends used for the reinforcement will have to be increased somewhat to reduce bearing stresses.

### 3.5.9 Prestressed concrete

Losses caused by creep and shrinkage will be higher than for normal weight concrete. This is an area that requires further study. Other design clauses may be modified in the same way as for reinforced concrete.

## References

1. Clarke, J. L. and Birjandi, F. K., Punching shear resistance of lightweight aggregate concrete slabs, *Magazine of Concrete Research*, **42** (152; September 1990), 171–6.
2. BS 8110: Structural use of concrete: Part 1, Code of practice for design and construction; Part 2, Code of practice for special circumstances, British Standards Institution, London, 1985.
3. BS 5400: Steel, concrete and composite bridges, Part 4, Code of practice for design of concrete bridges, British Standards Institution, London, 1990.
4. BS 8007: Design of concrete structures for retaining aqueous liquids, British Standards Institution, London, 1987.
5. BS 6349: Part 1, Maritime structures, British Standards Institution, London, 1984.
6. ACI 357.IR–91: State-of-the-art report on offshore concrete structures for the Arctic, American Concrete Institute, Detroit, May 1991, 118pp.
7. ACI 357.2R–88: State-of-the-art report on barge-like concrete structures, American Concrete Institute, Detroit, 1988, 89pp.
8. ACI 318–89 and ACI 318R–89: Building code requirements for reinforced concrete and Commentary, American Concrete Institute, Detroit, 1989.
9. NS 3473: Concrete structures, Norwegian Council for Building Standardization, 1989.
10. ENV 1992–1–1: Part 1: Design of concrete structures, Part 1 General rules and rules for buildings, Final draft, European Committee for Standardization, Brussels, 1991. (Part 1C, The use of lightweight concrete with closed structures, to be issued as ENV 1992–1–4.)
11. AS 3600: Concrete structures, Standards Association of Australia, Sydney, 1988.
12. Standard specification for design and construction of concrete structures. Part 1, Design, Japanese Society of Civil Engineers, Tokyo. (English translation, 1986.)
13. JASS 5: Japanese architectural standard specification for reinforced concrete work, Architectural Institute of Japan, Tokyo. (English translation, 1982.)
14. Mays, G. C. and Barnes, R. A., The performance of lightweight aggregate concrete structures in service, *The Structural Engineer*, **60** (No. 20; 15 October 1991), 351–61.
15. Institution of Structural Engineers/Concrete Society, *Guide to the Structural Use of Lightweight Aggregate Concrete*, London, 1987.
16. Forrest, J. C. M., An international review of the fire resistance of lightweight concrete, *International Journal of Lightweight Concrete*, **2** (No. 2; June 1989), 81–94.
17. Lawson, R. M., Fire resistance of ribbed concrete and composite slabs, *Concrete* (May 1987), 18–20.
18. Malhotra, H. L., *Spalling of Concrete in Fires*, Technical Note 118, Construction Industry Research and Information Association, London, 1984.
19. Ivey, D. L. and Buth, E., Shear capacity of lightweight concrete beams, *Journal of the American Concrete Institute*, **64** (No. 10; October 1967), 634–43.

20. Clarke, J. L., Shear strength of lightweight aggregate concrete beams, *Magazine of Concrete Research*, **39** (No. 141; 1987), 205–13.
21. Clarke, J. L. and Birjandi, F. K., The bond of reinforcement in lightweight aggregate concrete (to be published in *Magazine of Concrete Research*).
22. Hoff, A., Høiseth, K. and Haverstad, T. A., Testing of high strength lightweight aggregate concrete elements, *Nordic Concrete Research*, No. 3, Nordic Concrete Federation, Oslo, 1984, pp. 63–91.
23. Rowe, R. E. *et al.*, *Handbook to British Standard BS 8110: 1985, Structural Use of Concrete*, Palladium Publications, London, 1987.
24. Reynolds, G. C. *et al.*, *Shear Provisions for Prestressed Concrete in the Unified Code, CP 110: 1972*, Report 42.500, Cement and Concrete Association, London, 1974, 16pp.
25. Spratt, B. H., *The Structural Use of Lightweight Aggregate Concrete*, Report 45.023, Cement and Concrete Association, London, 1974.
26. Harrison, T. A., *Early-age Thermal Crack Control in Concrete*, Report 91, Construction Industry Research and Information, London, 1981.
27. Evans, D. J. and Clarke, J. L., *Thermal Movements in a Multi-storey Car Park*, Report 42.563, Cement and Concrete Association, London, April 1986.

# 4 Construction

R. N. W. PANKHURST

## 4.1 Introduction

### 4.1.1 Historical background

The use of lightweight aggregates, both coarse and fine, has been a feature of construction for centuries but, like most concreting materials, expectations of performance have risen, and now we are expecting a consistent, reliable material with predictable characteristics. The manufacturers meet these requirements, and those who could do so in a cost-effective manner have survived.

Thus, today we have available a range of coarse aggregates of varying densities capable of producing structural grade concrete of strength up to $60 \, \text{N/mm}^2$, while still showing substantial weight savings. The economics in use of these mixes can be shown to meet those of the cheaper, natural aggregate concretes provided the design of the structure from concept takes account of the economies available due to the reduced loads. Where there are long spans, where the self-weight to applied loads is high, then the lightweight aggregate concrete becomes substantially more cost-effective.

In the 1960s and 1970s, lightweight aggregate concrete was used in various structures, frequently with lightweight aggregate fines. Figure 4.1 shows on-site battery casting in 1965. It was a time of site-mixed concrete and there was little problem in adding additional bins behind the mixers to cater for these aggregates. With the increasing dominance of readymixed concrete in the 1970s the use of lightweight aggregates declined. They represented such a low incidence of work that it was not worth the effort of the readymixed supplier to add bins which would have relatively little use. It would only be offered at a substantial premium.

Also mitigating against lightweight aggregate concrete was the difficulty (almost, at the time, impossibility) of pumping it while pumping was becoming a significant means of placing concrete.

Thus the use of lightweight concrete became confined to specialist structures, such as stadia, bridges and other major works where site mixing was viable, or to structures with high dead to live load ratios where the high cost was offset by the very substantial advantages of the weight saving.

**Figure 4.1** University of East Anglia: battery casting on site with lightweight coarse and fines concrete, *circa* 1965.

The development of the thin light floor on profiled metal formwork, where the benefits of lightness and fire resistance made lightweight aggregate concrete almost essential, opened up its use and readymixed concrete plants installed the necessary storage to meet this demand.

Now in most major conurbations, there is a choice of suppliers with experienced technical expertise to provide lightweight aggregate concrete at competitive rates, although the use of lightweight aggregate fines is rare and still requires special facilities.

### 4.1.2 Lightweight aggregate in concrete

Whatever aggregates are used with cement and water to produce concrete they represent 70–80% of the weight of the materials used. Thus their influence is fundamental to the way the concrete performs in its unhardened as well as its hardened state.

Aggregates range from the hard, dense, crushed rocks through naturally occurring gravel aggregates to man-made lightweight aggregates. Even polystyrene aggregate is being manufactured and treated to provide lightweight concrete blocks.

The final outcome, hardened concrete, formed into load-bearing structures is required to have certain key properties. It needs to have strength in compression and it needs a certain amount of tensile strength, shear strength and an ability to bond to the other materials, particularly to steel reinforcement. It needs to be homogeneous so that it performs consistently through an element, to be consistent from one day to the next, and it needs to maintain this performance over many years.

It is obvious, therefore, that the aggregates make a major contribution to these requirements. Lightweight aggregates produce lightweight concrete, but because of the characteristics of the aggregate there are other important areas in which the performance of concrete is affected.

Lightweight aggregates are light due to the inclusion of air voids and it follows that they are absorbent, except for the very few with sealed cells. This absorbency plays an important part in the way the concrete performs in its wet state. Most lightweight aggregates are manufactured and hence are, by careful production control, uniform and consistent, which is important to mixing, placing and compaction.

Lightweight aggregates are weaker than natural aggregates and hence put some limitation on strength achieved by the concrete. This is, however, less significant because they are more compatible with the matrix allowing the whole to perform better in compression. Lightweight aggregates are generally inert and hence make little contribution to the chemical reactions which occur within concrete to the detriment of its durability. Sintered PFA aggregate adds some pozzolana to the concrete.

Lightweight aggregates often are made of waste materials occurring as by-products from other industries, and hence their use is ecologically desirable.

Lightweight aggregate concrete has a high strength/weight ratio. The aggregate is inert and not subject to alkali silica reaction, though the use of reactive sand could lead to reaction. It is pumpable, easily finished, easy to cut and drill, durable and is 'ecologically friendly'.

The contractor rarely has the opportunity to influence directly the decision to use lightweight concrete. Indirectly, however, both the contractor and the concrete supplier do influence the decision by their perception of the material and the difficulties they expect to result in its use. This affects the price they put upon its use, which in turn affects the design decision.

The designer in most cases makes a decision by balancing the cost against the advantages, such as weight saving, fire rating, etc. The costs are made up of the initial material unit cost and those applied by the supplier and the contractor, which may be loaded to take account of the additional risks they perceive in using a material unfamiliar to them and their operatives.

This chapter is directed towards correcting this distortion of the estimated costs of using lightweight aggregate concrete by giving a practical assessment of the facts to make realistic allowances on the use of lightweight aggregate.

Lightweight concrete has, for the contractor, both benefits and deficits. By knowing the characteristics of the material, full advantage can be taken of its potential benefits and the effects of its deficits can be minimised.

In lightweight concrete, the contractor not only has a material which is lighter, saving on cranage, but provided that correct procedures are followed, it is also easy to pump, place, compact and finish. It is consistent in strength, capable of achieving day-to-day strengths of up to $60 \text{N/mm}^2$ and which, by special measures, can reach over $70 \text{N/mm}^2$, giving very high strength/weight ratios. It does have a purchase cost premium and, if handled badly, will cause problems in pumping and finishing.

Success in the use of lightweight aggregate concrete will affect the economic assessment and hence the decisions of the designer to use it.

In this chapter the author, who has over 25 years' experience in the use of lightweight aggregate concrete, presents the reader with information to enable the avoidance of the pitfalls and the use of lightweight aggregate concrete with confidence, and thus avoid the need for 'loading' the price to cater for the unknown, allowing realistic pricing.

Much of the following information could, with appropriate changes, be applied to normal weight aggregate concrete, and had lightweight aggregate concrete been the norm then a similar chapter would be necessary to put forward information about normal weight concrete. Lightweight aggregate concrete is not difficult to use; it is different and methods need to be altered to take account of that difference.

## 4.2 Supply of lightweight aggregate

### 4.2.1 Bulk density and moisture content

Most lightweight aggregate manufacturers publish the average dry bulk density for their materials and this provides a satisfactory value for initial examination of mix design and costs. However, both the dry bulk density and the moisture content can alter, and unless these changes are measured and allowed for the consequences can affect both the costs and performance of the concrete.

Manufacturers should keep their customers aware of the bulk density of their materials, and this should be notified with the delivery tickets. It should also be allowed for in the loading if the materials are being sold by weight. The concrete producer must have this information to compute batch weights as the mix design is generally based on volume. Since most of these aggregates are the result of a hot manufacturing process, they are dry when transferred to the stock pile. Their moisture content at time of delivery will depend on the length of time in stock (hence on production and demand) and on the weather conditions. Thus the moisture content can vary over a broad range of values. The manufacturer should allow for moisture content when loading for delivery. The concrete producer also needs to measure the moisture content prior to batching as this must be allowed for in the batch weights of the materials and in the water added. Moisture content testing for lightweight aggregates is easily done using a 'speedy' meter since these aggregates are easily crushed, releasing the moisture. These tests for moisture content need to be updated throughout the day as variations can be considerable, potentially up to 15%.

### 4.2.2 *Controlling moisture content*

The ability of lightweight aggregates to absorb water is by far the most significant feature in their performance in concrete production and it is desirable to prevent this absorption during the concreting process. Some manufacturers soak the aggregates to an optimum level in their works and concrete producers will take steps to maintain the moisture in their own stock piles. This establishes a moisture level which remains consistent and predictable.

There are a number of methods used for soaking the aggregates. One is to immerse the aggregates, hot from their process, in water. This rapidly cools the entrapped air in the aggregates which reduces in volume drawing water into the aggregate. This is effective with some aggregates with fairly open pores, but less effective with those made of sintered pulverised-fuel ash (PFA) such as Lytag.

Continuous sprinkling of the stock piles both in the manufacturer's yard and in the concrete plant will again introduce sufficient moisture to reach equilibrium in the open-textured materials. It is important that the pile be free-draining, to ensure consistency. This method is not entirely satisfactory but is commonly used.

A development of this process is to contain the water and aggregate in a 'lagoon', agitating the aggregate. This is the most effective method at atmospheric pressures but is still not totally effective, particularly for the closer cell materials, such as sintered PFA.

The most effective method, applicable to **all** lightweight aggregates is vacuum-soaking. The aggregates are introduced into a pressure vessel and a vacuum of 60–80% is created. The aggregates are then immersed in water maintaining the vacuum with some agitation. After a short time the vacuum is released. This can result in almost full absorption. Figure 4.2 shows typical equipment.

**Figure 4.2** Trial vacuum-soaking equipment.

Although vacuum-soaking is expensive it is the only fully satisfactory method for sintered PFA and other fine cell materials. The material when soaked must be continuously water sprayed until batched to minimise drying out.

Unless the aggregates are nearly saturated when batched, then inevitably absorption occurs during mixing which must be allowed for in the amount of water added. The problem is exacerbated by pumping when high pressures are applied, forcing variable amounts of the water into the aggregates to the consequent detriment of the workability. (This is discussed later.) Fortunately the water absorbed has no detrimental effect on strength or durability and so can and must be ignored when calculating the water/cement ratio. Non-saturated aggregates require increased mixing times to reach equilibrium.

## 4.3  Mix designs

### 4.3.1  Introduction

Lightweight aggregates are weaker than normal weight aggregates and in order to compensate for this higher cement contents are used, while the water/cement ratio is kept low by water-reducing admixtures. Even the use of these richer mortars would not produce the strengths achieved without other advantages found in lightweight aggregates. These are their ability to bond to the cement matrix and the closeness of their coefficients of thermal expansion to that of the matrix, which leads to a very coherent concrete with little or no microcracking. The water absorbed also aids extended hydration.

Other advantages of this compatibility between the aggregate and matrix is the reduction of permeability leading to improved durability and a reduced tendency to shrinkage cracking, probably owing to the lack of microcracks acting as stress raisers.

The manufacturers all provide a range of recommended mix designs for the strengths required in the majority of structures both for placing by crane or pump. These mixes have been developed from long-term experience and may need adjustment only if the sand or cement being used is unusual. Often a harsh sand may cause bleeding or pumping problems, and this must be compensated for. With expedience, the concrete producer too will make fine adjustments to suit the materials, circumstances, experience and controls.

Mix designs are also available for concrete mixes using partial cement replacement by PFA or ground-granulated blastfurnace slag.

As with normal weight concretes, these materials provide advantages. They not only have the benefits of economy, but pumpability is improved and the mix is even more resistant to bleeding and thermal cracking. Particularly for lightweight pumping, these replacements allow simpler, cheaper admixtures to be used by improved particle size distribution of the mortar.

### 4.3.2   Lightweight fines

Generally for structural concrete the lightweight aggregate is confined to the coarse faction. This gives a substantial weight saving with the least increase in cost.

Where even lower densities are required and the strength requirements preclude the use of the lighter aggregates, the fines too can be lightweight. These fines, like the coarse material, have absorbency and they also include very fine material, increasing water demand further. Lightweight fines are difficult to handle and batch, adhering to bins and weigh gear. The consequences to the performance of the wet concrete is to make it 'sticky' and liable to substantial loss of workability as absorbency takes place. Pumping, too, is more difficult and trial mixes are recommended to establish the amount of water necessary to cater for the absorption.

It is desirable and possible to overcome most of this effect during an extended mixing cycle.

### 4.3.3   Pumped concrete

Pumping is widely used for placing lightweight aggregate concrete, as illustrated in Figure 4.3. Mix designs for pumping need to take greater

**Figure 4.3** Pumped lightweight aggregate concrete being placed on metal decking.

account of the absorption likely to occur under pressure. The interdependence of pumping, site conditions and materials require this subject to be covered in the section on pumping. Pump mix designs have evolved to meet the changing demands of the industry and the improvements in pump designs, and although the manufacturers have provided their recommendations these mixes need adjustment to meet the requirement of the particular use, and materials used, more than for any other method of handling.

When the mix is to be pumped, then the use of admixtures, length and height of pipeline, type of pump and pump pressures have to be taken into account.

For extreme pumping conditions, as will be shown later, trials are advisable to ensure that the mix performance will be optimum.

### 4.3.4 Mix designs for pumping

Mix designs for pumping need to take greater account of the absorption which is exaggerated by the pressures generated in the pipeline. There has been much development in dealing with this problem which varies very much between types of aggregate. For mixes consisting of lightweight coarse aggregate, sand and ordinary Portland cement, it is necessary to use a thickening agent to reduce the movement of the water from the matrix to the aggregate and to then add a plasticiser to recover the workability. A similar approach is made in the case of lightweight coarse and fines. Trials are recommended to establish the dosage and water content.

When PFA or ground-granulated blastfurnace cement are used, these provide enough added fines to make the thickening admixture unnecessary, enabling the use of a plasticiser alone to suffice, and this is adequate for most normal site requirements. Thus the use of cement replacements are very cost-effective.

Both the aggregate supplier and the concrete supplier will usually have sufficient experience and records to produce proven mix designs, but if there are to be either materials, site conditions, strength or workability requirements which are untried, then trials must be carried out to give confidence of a satisfactory performance.

Mix designs for pumping concrete made with both coarse and fine lightweight aggregate can be achieved. The strength levels attainable are less than with sand fines but $40\,N/mm^2$ is achievable on a regular basis.

The workability needs to be of a high order, about $650-700\,mm$ flow, and the cement content increased by some $70-100\,kg/m^3$ above the equivalent sand based mix. Obviously this means an expensive material which is only used when it is required for a particular purpose.

## 4.4 Batching

### 4.4.1 Aggregate proportion

As has been discussed, the lightweight aggregate received by the mixing plant can vary in bulk density owing to two factors. One is the variable dry bulk density of the lightweight material itself and the other is the amount of water absorbed in it.

The supplier should give notification of any changes in density and the concrete producer will have carried out moisture content tests and bulk density tests to allow these to be taken into account in the batch quantities. Failure to do so would affect the strength, workability, pumpability, final density and yield of the concrete produced.

Most manufacturers state a bulk density in their literature. This is only good enough for estimating costs and mix designs but should not be relied on further without confirmation of its current validity.

Account of the moisture must be taken, not only for its effect on batch weights of the aggregates, but for its contribution to the water content of the mix. There needs to be provision in the batched water for bringing this aggregate moisture content to the anticipated level of water absorbed at atmospheric pressure (up to 12% for sintered PFA). This value will be established by experience.

It is important for the mixing plant to have good communications with the site so that the site informs the plant of the condition of the mix they receive to enable batch water to be adjusted to cater for changes en route.

The mix proportions set by the design are changed only by the adjustments required to take account of density changes and moisture content, unless observation shows that the character of the mix itself is unsuitable. The water added, although it can be estimated by taking these factors into account, is not accurately predictable because of the absorption. The final test for water content is the workability of the concrete. This applies with any concrete, even when absorbent aggregate is not used. It is the free water that is being specified; particularly in the case of lightweight aggregate the workability is sensitive to small changes in free water and this is an accurate and reliable indicator of the water/cement ratio. If the workability can be maintained within the design limits, the performance of the concrete is assured. Thus the batching of water is checked against the resulting workability rather than computed values, although these are used as a starting point.

This principle is used throughout the chain of processes from mixer to placing, and will again be referred to in section 4.5 on pumping.

### 4.4.2  Mixing

Mixing procedures for lightweight aggregate concrete should be adjusted to take account of the absorption of the aggregates and to avoid 'balling' (the forming of balls of semi-dry material which do not break up but roll round the mixer paddles or truck mixer blades).

To take account of the absorption, the workability has to be reassessed at the end of each process and, if necessary, the water increased to replace the absorbed water and to allow for the further absorption expected during delivery to site.

Balling, which occurs most commonly when lightweight fines are used or when cement replacements are included in the mix, can to a great extent be avoided. It is caused by much the same reasons as lumpiness in cocoa when mixed with too much milk. The method for prevention is similar. The materials must be first mixed with a much reduced water content – enough only to produce a slump of about 50 mm. When these materials have been thoroughly mixed the remainder of the water is added. There is some temptation for the batching plant operators to short cut this process when under pressure, but this should be resisted as the consequence of balled concrete arriving on site is to invite rejection.

Mixing concrete containing both lightweight coarse and fine aggregates presents no special difficulties provided that the mixing sequence is set up to avoid balling. In the sand-based mixes absorption occurs, but now the fine materials are also absorbing water. Fortunately, because it is of small particle size, this absorption is easy and relatively short lived; providing the mixing time is sufficiently long to cover the period of absorption then further absorption by the fines no longer affects the workability and the rules for a sand-based mix apply.

A simple but successful means of establishing the necessary mixing time is to place a quantity of the mixed aggregates in a drum and fill this with water. The aggregate is agitated and as absorption takes place the water level will fall. The time taken before the water ceases to drop is an indication of the time necessary to mix the materials.

The procedure of adding water at site is looked upon with distrust by supervising engineers, but is a perfectly valid process allowed for in BS 5328 as long as it is supervised and controlled to an agreed procedure, as given in the appendix at the end of this chapter.

It is a practice not confined to the foregoing situation but is applicable for any concrete. It is particularly relevant to lightweight aggregate concrete delivered to site for pumping.

For lightweight concrete, as for normal weight concrete, good practice should be followed. Water may be added under strict control to bring the workability to the intended and designed level – under the supervision of a technician and certified by testing and recording. The pump driver should never be allowed to just put a hose to the hopper. This way leads to anarchy and a whole range of problems from segregation in the line with blockages to low strengths, drying shrinkage cracks and plastic settlement.

### 4.4.3  Yield

There is often dispute between the supplier of the concrete and the user concerning quantity. This can be a problem for any type of concrete and it is as well to understand the causes and their solution to avoid disputes.

Concrete is batched by weight but its use is measured by volume. A floor is not placed at $200 \, kg/m^2$ but is 100 mm thick. Thus, if the concrete supplier overestimates yield the contractor will complain of short supply. The problem is not always caused by the concrete supplier. The contractor often does not accurately measure the pour, or does not take account of the deflection of formwork, particularly of profiled metal formwork which can increase the volume by as much as 10%. These contractor's errors of measurement can be cured by greater care.

For the concrete supplier the cause of yield variations once more arises from the variability in density of the aggregate owing to changes in the aggregate itself and to moisture content. To avoid errors these properties must be measured regularly and the batch weights adjusted accordingly.

When supplying pumpable lightweight concrete the workability is such that it is not possible for truck mixers to travel fully laden. If they do there is almost certainly going to be spillage from the drum when manoeuvring on the road – accelerating from intersections can result in the following traffic receiving part of the load! This obviously is unacceptable to other road users and affects the quantity of material that reaches the site. It is normal to run pumpable lightweight concrete in reduced loads, typically reducing by about 10%. An alternative to reducing the load is to reduce the workability for the delivery and allow the necessary water at site to bring the mix to a pumpable quality.

When reduced loads are used to avoid spillage of pumpable concrete, failure of the shipper to adjust the recorded quantity on the delivery ticket sometimes occurs, which obviously causes disputes with the contractor. Both parties should be aware of the potential for problems of yield and should take care that they get their part of the operation right to avoid unnecessary disputes.

## 4.5  Pumping

### 4.5.1  Developments in pumping practice

Pumping lightweight aggregate concrete is the area where its character has the most impact on conventional methods.

As has been discussed, lightweight aggregates have the characteristic of absorbency, and in most areas of work this occurs at atmospheric pressure so that it is easily simulated, and by the time it leaves the mixing plant a level of equilibrium has been reached. There may be some further absorption on delivery to site, particularly if there has been a delay in hot weather or if the aggregates were particularly dry in the bins, in which case some water should be added at site under approved procedures (as discussed in more detail earlier) to recover the lost workability.

When pumping, however, there is a substantial increase in pressure and this compresses the air within the aggregate allowing further absorption to occur, causing workability changes. Providing this is overcome, pumping becomes easy with high outputs possible and the lower weight of the material reducing pressures in high-rise lines.

It has taken several years and much development to reach this stage in the technique of pumping concrete generally and lightweight concrete in particular. In the 1960s and early 1970s pumping lightweight concrete was considered to be unsatisfactory, if not impossible. Most pumps were reciprocating piston pumps using crude valve assemblies. These valves allowed some of the concrete to be taken back into the cylinder at each stroke – double working it, which is to be avoided – and it was not possible to maintain a watertight seal around these valves. Any loss of water from the matrix compounded the effect of absorption, quickly leading to blockages. The shape and nature of the valve assembly caused severe changes in shape and section of the pipe through which the concrete had to pass, and this resulted in high local pressure which, even if not causing a blockage, could overload the pump mechanism. Lightweight aggregate gradings were changed and blended to give a more continuous grading specifically for pumping.

There were pumps which pumped lightweight aggregate concrete satisfactorily. These were the peristaltic pumps, where the concrete is pumped through a flexible hose by rollers which squeeze the hose while moving along it, forcing the concrete ahead of them. Unfortunately, at that time the hoses only lasted for some $500\,\text{m}^3$. These pumps have now been

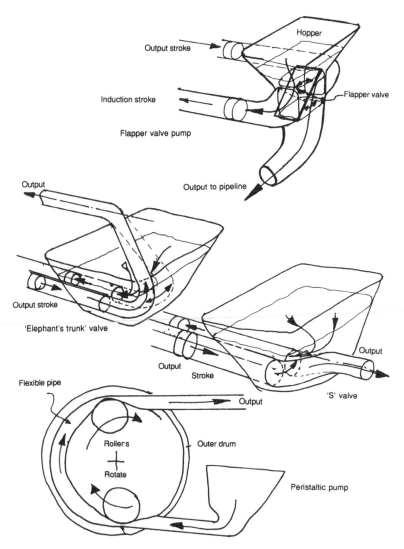

**Figure 4.4** Typical concrete pumping systems.

developed to a much higher standard and can hold their own against most of the pump systems available.

Another type of pump that was capable of pumping lightweight concrete was the compressed air pump, or 'placer'. This had a container into which the concrete was placed and the container then sealed and connected to a compressed air reservoir forcing the concrete into a discharge hose. Although effective, it had limited output and as the air pressures were high it was dangerous should there be a failure.

A milestone in the development of pumping lightweight concrete came in the early 1970s when the chemical admixture industry produced the 'pump aid'. This was a thickening agent which inhibited the free flow of water through the matrix containing it, reducing its absorption into the aggregate. This was combined with a water-reducing agent – a super-plasticiser to recover the workability. Pumping with the flapper valve pump now became successful for limited pipelines – 5 in (125 mm) lines were advisable and lengths of 50 m were considered acceptable.

Then came the development of the straight-through valve such as the 'elephant's trunk' and 'S' valve, where the pipeline was connected to a curved tube which oscillated between two cylinders. Its end slid across a plate called a 'spectacle' plate because it had two holes in it which were the cylinder openings. When the pipe was over a cylinder there was a straight-through smooth passage for the concrete and the other cylinder was directly exposed to the concrete in the hopper within which the system was submerged. Using these pumps and pump aid, pipelines of 100 m are easily catered for and heights of 50–60 m are commonplace. Figure 4.4 illustrates the various systems that have been discussed.

The use of cement replacements has enabled simpler admixtures to be used while still providing very pumpable concrete. They did not need a thickening agent and an ordinary plasticiser sufficed.

### 4.5.2 Pumping for high-rise buildings

These developments coincided with the growth of high-rise steel-framed structures using lightweight concrete floors. There was no crane access and pumping was the only viable means of placing. In the United States, by saturation of their aggregates (which was easier than with the UK materials), pumping 300 m high was achieved. In the UK, 50–70 m was the limit.

Experience on some projects with long horizontal lines, up to 200 m and 50 m risers, showed the significance of the difference in aggregate proportions. If the workability was above 650 mm on the flow table it was no solution to pipeline blockages to increase the workability if the materials were badly graded or proportioned. To overcome pumping problems in long lines, water added to increase workability may just

**Figure 4.5** Canary Wharf Tower.

cause segregation and bleeding, whereas by minor changes to the mix
proportions – increasing the cement and sand – the pumpability can be
dramatically changed without increasing the flow.

### 4.5.3   Canary Wharf trials and experience

A quantum leap occurred in the pumping of lightweight aggregate concrete
in the UK when the Canary Wharf Tower (see Figure 4.5) was constructed.
This was very much higher than any previous UK structure and four times
the height of the highest lightweight pump previously achieved in the UK.

The contractor and ready-mixed concrete supplier, Ellis-Don McAlpine
and Ready Mixed Concrete (London) Ltd, were faced with a requirement
beyond their experience. The cooperation between these organisations
early in the contract achieved a satisfactory method and provides a good
example of how good preparation and testing prevents difficulties during
the contract.

The options for getting the concrete to over 250 m above the mixing
plant were by crane or hoist to a high level pump, by stage pumping, or
to use a single pump and riser for the whole height. There were objections

to the first three options. At each pump stage some reworking of the mix would be necessary and the controls and coordination made them laborious and unsatisfactory, whereas once the principle of a single pipeline and pump had been proved it would be familiar ground.

In the United States it was found that, providing full saturation of the aggregates had been achieved, the lightweight concrete pumped with no greater difficulty than normal weight concrete. It was, therefore, obvious that the first step to investigate was ways and means for achieving saturated aggregate. Previous attempts to soak Lytag, the chosen aggregate (the only readily available lightweight aggregate in the area), at atmospheric pressures proved unsuccessful, so once more looking to the American experience it was decided to use vacuum-soaking.

The mixing plant to be used on this contract came from Norway where it was being used for the construction of oil rig platforms for the North Sea. There was in Norway experience in lightweight aggregate concrete and pumping and very large powerful pumps were in use. Therefore, Ready Mixed Concrete (London) Ltd decided to carry out full scale trials in Norway.

A suitable cylinder was obtained and adapted for the vacumm-soaking process. A test pipeline was set up simulating the resistance likely to be experienced at Canary Wharf by using a long horizontal line made up from a number of parallel pipes connected by short radius bends and ending in a vertical line of 52 m height set into a tower crane mast. Although the height of the discharge was not great, experience had shown that the long horizontal line would provide the equivalent resistance of the vertical line to be used at Canary Wharf. In the view of the Norwegians the frictional resistance of a horizontal line equated to the hydraulic pressure due to the head of concrete in the riser. A relatively short riser was put at the end to produce sufficient head to load the system.

Using the vacuum retort the aggregate was loaded in batches. A vacuum of 80% was applied and maintained for 30 minutes and then water was added while agitation of the aggregate took place. The vacuum was released and the aggregate discharged to a bin where it was constantly sprinkled until used. The trials showed that this process produced an aggregate that had excellent pumping properties and could be pumped to the top of the Canary Wharf Tower without any significant loss of workability.

The set up used in Dockland was driven by a large Schwing pump driving a 5 in (125 mm) pipeline. The first 30 m section of line was horizontal. It then entered a long radius bend upwards to join the riser, which had another long radius bend at its top with horizontal distribution line with a maximum of two 3 m flexible hoses. The riser extended as the work rose.

Pump manufacturers often recommend that for high-rise lines the riser is broken at intervals by two long radius bends and a short horizontal straight. They argue that this reduces the back pressure at rest.

The Canary Wharf team felt that the benefits of this arrangement were uncertain and did not offset the disadvantages of the friction of the added bends. Their riser was unbroken bottom to top. In use there appeared no handicap to this decision but one practical advantage. If there were a blockage it was fairly sure to be at this one bend at the bottom of the riser. Hunting for blockages all the way up 250 m of riser, a time-wasting chore, was thus avoided. They did have cause to appreciate this decision as there was a blockage at the bend only once (a credit to their planning), and it was easily cleared. They then learned that several hundred feet head of concrete, when suddenly released at a blockage, has a lot of energy and potential for making a lot of mess – about a 3 m cube of concrete was ejected at high velocity under a pressure of $4\,N/mm^2$ (nearly $600\,lb/in^2$)!

During the construction of Canary Wharf Tower the site team carried out one pour at level 42 using aggregate that had not been vacuum-soaked although it had been immersed in water for three days to achieve 16% moisture. This mix pumped without blockage but the pressures were higher. Experience elsewhere has been that when 100 m or more of pipeline is used a near optimum of forced absorption has occurred and further increase in pressure and time makes little significant difference. The author's experience has been so, having pumped unsoaked Lytag aggregate concrete through 250 m of line, of which some 60 m was vertical.

Once the mix has been adjusted to provide adequate workability without segregation – achieved by slightly increasing the water, cement and sand – then the performance from a rise of 10–60 m made little difference to the pump pressures and output. Before these adjustments were made, even 100 m of line was causing difficulties, and increasing water alone only led to segregation.

### 4.5.4 Recommendations for pumping

Very high pressures are generated in concrete pump pipelines and, when failures occur, they are dangerous. Pipes must be regularly inspected for wear, particularly at bends. Clips should be undamaged, in good condition and properly attached, and if compressed air is used when cleaning, the discharge should have a baffle to protect the operatives. Rubber pipes must never be used anywhere other than at the discharge end of the line.

When setting up the pipeline, use long radius bends and 125 mm line wherever possible. If reducers are used they must be gradual.

The bend at the bottom of a riser should be encased in concrete to control the high loads acting, and the pump should also be restrained. Grouting the line prior to pumping is important and with long lines it is an advantage to have intermediate grouting hoppers along the line. Using good pumping practice and ensuring that the mix design is adjusted to meet the needs of the materials used, pumping lightweight aggregate concrete is easily achievable.

Vacuum-soaking is expensive and only appropriate to large-scale projects. For typical sized work, adjustments of the mix proportions, possibly more cement and increased admixtures, could meet the requirements at a modest increase in cost and slightly higher pump pressure. It would have a lower margin of safety before blocking but it would do.

When pumping lightweight aggregate concrete that has not been vacuum soaked, there is a loss of workability in the pipeline. This is greater the lower the moisture content of the aggregates at mixing. Also, it is greater if the workability at the pump is near the lower limits – say, 550–600 mm flow. At 650–700 mm flow the change is negligible.

When sampling for cube strengths the accepted practice is to sample at the discharge from the mixer into the pump hopper. It is arguable that sampling at the pump discharge more closely represents the concrete in the work. Experience shows that a 2–5 N/mm$^2$ advantage in cube strength is found by doing this.

As has been discussed elsewhere, it is often necessary to adjust the workability of truck-mixed concrete before discharge at site; when pumping, this facility is essential. Good control at the batching plant and communication between the site and the plant can minimise this, but not altogether remove the need. If this facility is not available, then there will be loads that will not pump, and these will be rejected as they are too dry. Alternatively, there will be loads where the batcher has added too much water to avoid drying, and these will also be rejected as they are too wet. The consequence will be gaps in delivery leading to concrete being in the line too long, which will lead to blockages.

Before any pour starts, a procedure for adding water at site, under close supervision, must be agreed between the contractor, supervising engineer and supplier. There needs to be an agreed written procedure, including a table indicating the amount of water allowed to be added for various flow levels.

Although it has been recommended that 125 mm pipelines are used, it is possible and sometimes very convenient to use 2.5 in (60 mm) lines. Typical occasions are small infill pours and 'come back' works where cranage is unavailable. The usual options are to use a hoist and barrows or on-site mixing of pre-bagged materials. Both are unsatisfactory and expensive.

Using a high-pressure pump of the type used for shotcrete and cement-rich mixes of high workability, pumping in 2.5 in lines is quite satisfactory. The output is low and pressures high but for the examples given it is an economic solution, simple to set up and dismantle as the pipelines are light and easily handled.

## 4.6 Placing, compaction and finishes

### 4.6.1 Formed finishes

Lightweight aggregate concrete, because of its high cement content, is capable of high-quality formed finishes. Owing to the regular and relatively small coarse aggregate there is no tendency for 'aggregate lock'. As with similar normal weight concrete the use of impervious smooth moulds increases the occurrence of air bubbles on the surface. This is even more likely when using lightweight aggregate fines. As with normal weight concrete, this effect can be greatly reduced by the removal of the polish from the forms. Steel moulds particularly can be acid etched. This reduces both the incidence of air voids and unevenness of surface colouring.

Bleeding is rarely a problem except in lightweight concrete designed for pumping which will bleed when, either by mistake or design, there is insufficient fines and too much water. This can easily be corrected by adjustments to the mix.

Overvibration of lightweight aggregate concrete, unlike normal weight concrete, does not cause the concentration of fines and cement at the top of a pour with its consequent shrinkage cracking and 'scabbing'. Instead, the coarse aggregate rises and, at worse, a layer of coarse aggregate floats above the surface creating a loose layer. This is easily observed and removed, exposing the dense concrete below.

Because lightweight aggregate concrete is more tolerant to strain it can be cast in larger bays. Wall pours and slab pours can safely be cast in larger lengths. In the case of water-retaining structures, the risk of cracking is substantially reduced.

### 4.6.2 Floor slabs

When casting on profile metal formwork the benefits of lightweight aggregate concrete can be fully appreciated. In these circumstances, because of the character of the concrete and the strain distribution by the metal formwork and steel framework, each bay contains its strain. Pour sizes are governed only by economics and the resources which can be applied to the work. Pours of $1000 \, m^2$ are commonplace, and pours are on record in excess of $3000 \, m^2$ in a single shift. Obviously the organisation of delivery of concrete and the finishing operations must be carefully considered for work on this scale.

The concrete can be spread and levelled using rakes and broad-bladed spreaders and any of the proprietary vibrating beams or 'razor backs'.

When placing concrete on profile metal formwork, the vibration from the vibrating levelling systems is very effective and the width of bay is determined only by the column spacing.

Pumpable lightweight aggregate is of a near flowing consistency and

**Figure 4.6** Lightweight aggregate risen to the surface.

needs very little vibration. The author has carried out tests on 600 mm flow Lytag concrete, where test cubes were made without compaction, and the strength compared with that of cubes from the same sample compacted in layers in accordance with the code of practice. There was no significant difference in strength. However, it would be unwise to rely on this as justification for not vibrating, but it does indicate the ease with which the material can be compacted.

### 4.6.3 Unformed finishes

This is an area where lightweight aggregate concrete has a particular character requiring special consideration. Because the aggregate is lighter than the matrix it will float and rise within the mix. This will continue to happen until the concrete reaches its initial set. This effect creates a bubbly appearance to any exposed surface, as illustrated in Figure 4.6. It affects most fine finishes which are created when the concrete is fluid.

Of the most common finishes – rough tamp, fine tamp, brushed, trowelled and power floated – all except power floated are affected to some degree, although rough tamp masks the effect. Power floating, however, produces a hard smooth finish of high quality.

### 4.6.4 Power floating

When finishing lightweight concrete which has been delivered by pump, inconsistencies in water content and aggregate moisture content have an important effect on the whole operation. Excess water leads to bleeding which tends to retard the set – probably influenced further by the admixtures used. The variability of setting times are particularly important when power floating. In extreme cases parts of the slab will be passing the point suited to power floating while other areas will still be too soft to take the power float. Consistency of workability should always be the objective.

In cold weather the set may be delayed for over 10 hours before power floating can be completed. This creates problems of protection against frost. Screening against wind and heating from below are sometimes necessary. Heated concrete will do much to mitigate against this problem.

There have been attempts to retard the set deliberately to overcome the extended working hours caused by slow setting, bringing the finishing into the following day shift. This is difficult to achieve with sufficient consistency to give confidence to leave the work. (Should set occur overnight the finish is irrecoverable).

### 4.6.5   Computer floors

One area of difficulty which has recently come to light is the problem of adhesion of computer floor jacks to the top of a power floated lightweight concrete floor. This is a very common circumstance although the need to have such a slab power floated when it is only to be covered is arguable. The difficulty came to light when the 'sandbag' test popularised by the Property Services Agency in the UK was used. It was found that this caused the jacks to pull away the top surface of the concrete. A cure for this was to lightly scabble or scarify the surface at the position of the jack when the bond was complete. A procedure adopted by some contractors was to use a light scabbler to set out for the jacks instead of the traditional dab of paint.

### 4.6.6   Weather

Weather conditions affect lightweight concrete just as they do normal weight, but there are certain areas where the effects are accentuated. Unless wind breaks are provided, warm drying winds remove surface water and cause the surface of the concrete to set early. Because the high fines content reduces water movement this is confined to the surface. The effect can create a hard skin which, when over a pumped concrete, will be ready for power floating, but the underlying concrete will not take the weight of the equipment. Protection against surface drying is important for all concrete. Lightweight aggregate does, however, have the absorbed water to maintain curing even when inadequate precautions have been taken.

Cold weather must be protected against, although the voids in the aggregates provide relief to the expansion of the water when freezing. The insulating effect of the aggregate gives a measure of protection to the bulk of the work. The surface, however, can suffer superficial frost damage as easily as can normal weight concrete, and the same precautions should be observed. Particular care should be taken when casting thin sections on profiled metal formwork. There is very little volume and no insulation as is normally provided by timber forms beneath, so that there is a very high surface/volume ratio. The underside needs to be protected against cold and, in subzero temperatures, should be heated during curing.

Once the concrete has reached a strength of $10\,\text{N/mm}^2$ the risks are minimised provided that the temperature fall in the concrete is at a rate which allows the pressure to be relieved.

The author would draw the readers' attention to the effect of the use of salts to disperse snow and ice. This should never be done on either lightweight or dense concrete in the first months of ageing and should be avoided on non air-entrained concrete. The effect on even hard, dense or lightweight concrete can be devastating when saturated. The heat lost to the melting of the snow or ice is so great and so rapid, it can take only seconds to drop over 20°C, causing the water in the upper layers of the concrete to freeze very rapidly, and 5–10 mm of the surface will delaminate over the entire area, as shown in Figure 4.7. Once the concrete has dried out, this effect is minimised and the air voids in the lightweight aggregate concrete act, as do those in air-entrained concrete, to relieve the pressure.

There is also the additional risk to reinforcement owing to reduced cover, and from the chloride in the saline solution. This danger is common to both normal and lightweight concrete.

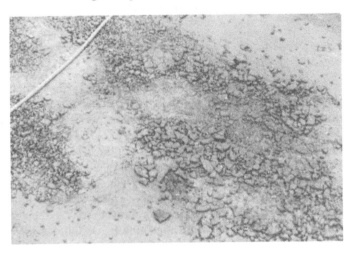

**Figure 4.7** Frost damage to saturated concrete.

### 4.6.7 Vacuum de-watering

There is some concern that lightweight aggregate concrete, because of the water contained within the aggregate, will create problems when used in roof slabs when a waterproof membrane is applied to the top. This is thought to be even more of a problem when cast on profiled metal formwork (which is also impervious to water vapour) using pumpable lightweight concrete. Attempts to overcome this problem have been made by drilling thousands of holes in the metal formwork and/or by using vacuum de-watering of the concrete when laid. Drilling holes in the metal deck provides such small areas for evaporation that it is pointless.

Vacuum de-watering of lightweight aggregate concrete can be done but it needs great care in ensuring the complete seal around the blankets. Even at best it does not remove sufficient water to overcome the perceived problem. The nature of the concrete, its high fines content and the fact that much of the water targeted is within the aggregate, prevents this method removing sufficient water. It does, however, provide benefits to the use of power floating, enabling that operation to occur earlier and at a predictable time, but these are offset by the restricted bay sizes necessary and the very much reduced production rates.

Lightweight aggregate concrete will dry out in time. Good practice dictates a waterproofing system with provision for dispersion and venting of vapour trapped beneath it.

## 4.7 Testing lightweight aggregate concrete

### 4.7.1 Strength

The same tests apply to both lightweight aggregate concrete and normal weight concrete. The choice of tests and their relative importance differ. Strength tests are unaltered and the cube, cylinder or beam tests are carried out in the normal manner and their results used in the normal way.

### 4.7.2 Workability

Workability tests, although carried out in the normal manner, may need to be interpreted differently. The slump test tends to underestimate the workability of the lightweight aggregate concrete. The user will soon learn from experience how to modify his judgement.

The required workability of lightweight aggregate concrete for pumping is very much greater than is normal weight concrete – indeed, if measured by slump, the values will exceed 200 mm! At this level the slump test is insufficiently sensitive to be useful, and one must use the flow table test. Flows of over 550 mm are necessary, up to as high as 700 mm.

As a 160 mm slump is considered to equate to a 480 mm flow, it can be seen that for flows of over 600 mm the slump will be fully collapsed. The flow table test is recognised internationally and is described in BS 1881.

To the uninitiated, the visual judgement of the mix by its appearance will be totally misleading. What appears to be excessively workable by observation will, when tested on the flow table, prove too dry to pump! The author, who has been working with concrete for nearly 40 years and with lightweight aggregate concrete for over 25 years, would not trust his judgement without confirmation by flow table testing. However, a fairly good 'rule of thumb' for judging pumpability for all except extreme circumstances is to observe the paddles passing into the surface of the concrete in the hopper of the pump. If the paddles pass in without leaving a depression in the surface, then the mix will not be too dry – obviously it may be too wet!

### 4.7.3   Testing for density

There are several density characteristics of concrete.

1. The fresh wet density applicable to concrete at or before point of placing.
2. The saturated density of the hardened concrete usually applied to test cubes or cylinders straight out of the curing tanks.
3. The air-dried density, usually representing concrete *in-situ* or samples which have been allowed to dry naturally to a level of equilibrium, which represents a sensible basis for dead weight design in the permanent structure.
4. The oven-dried density which is the density of samples which have been treated in an oven until all moisture has evaporated and no further weight loss occurs.

Of these values, the fresh wet, saturated and oven-dried are specific, easily achieved and unambiguous.

Air dried can vary with atmospheric conditions and will be different from concrete in an air-conditioned building, a swimming pool, a bridge in a hot arid climate, or in a wet humid climate. It is, however, the permanent work designers target, and he needs to define what conditions need to be met.

For the specifier there is a need to have a measurement to predict the density, which allows a result to be inspected before any commitment of use is made. Fresh wet density can be used, but it must be in the knowledge that this measurement varies with the absorbed moisture content of the aggregate.

The saturated or oven-dried are also specific conditions which the specifier and the constructor can agree on. From these, most lightweight aggregate suppliers can advise on the adjustment to predict the air-dried density. As a rule of thumb, the final air-dried density equates to the oven-dried density plus $50 \, kg/m^3$. The specifier must realise that these densities vary and must ensure that he does not set too narrow a target – the supplier of the aggregate will be able to predict the average and the range of density within which the concrete will fall. It is difficult, indeed impossible, to change the characteristic densities of the available lightweight aggregate concrete. It is thus the duty of the designer to design within the limits of the material available.

### 4.7.4  In-situ strength testing

Cores, of course, perform much as cubes do. Other methods of test, such as Schmidt Hammer, 'Pundit', Windsor Probe should be used as comparative tests to measure against known similar concrete of proved strength so that the particular character of the lightweight aggregate is taken into account.

The CAPO & LOK tests, which can be used with great confidence, do differ in performance in lightweight aggregate concrete. However, the manufacturers provide calibration graphs to suit each material and by using these the results are found to be accurate, consistent and with a high confidence level.

### 4.7.5  Performance in fire

Although not strictly the responsibility of the contractor, the performance of lightweight concrete in fires is worthy of comment as sometimes, as has happened recently, the fires occur during the final stage of the construction phase. Site hutting and materials are notoriously vulnerable to fire, and as these fires occur before the sprinkler system is in place, the period and intensity of the fire can be severe.

Experience has shown that in these circumstances on site, *in-situ* lightweight concrete lives up to the reputation achieved in the many laboratory demonstrations. On one major project which suffered such a fire, the lightweight concrete slab, only 130 mm thick overall (70 mm between the ribs), cast on profiled metal formwork, maintained its integrity even though the supporting structural steel beams deformed and deflected as much as 900 mm. The lightweight concrete thus confined the blaze to one floor and prevented much more serious consequences. Part of the slab was the infill of crane openings and was only 48 hours old. There was no evidence of the damage that was suspected would happen in such a case owing to the expansion of the absorbed water into steam.

### 4.7.6 *Fixing into lightweight aggregate concrete*

For concrete using lightweight aggregate, cast-in anchors generally perform as in normal weight concrete. In the case of post-fixed anchors, some types may have to be downrated, although the under-reamed types have been found to be equally effective in lightweight aggregate concrete. The advice and recommendations of the manufacturer should be sought and followed. When in doubt, proof testing should be used to give assurance.

### 4.7.7 *Making good lightweight aggregate concrete*

The success of making good required that the new material must be durable and must match in character the 'host' material in strength and thermal expansion. It must also, of course, have a strong permanent adhesion to the host.

Most conventional making-good materials qualify in their compatibility with the lightweight aggregate concrete. Durability and adhesion are, as always, dependent on following good practice. There has to be careful preparation of the area to be made good, followed by adequate curing. The manufacturer's instructions should always be followed.

### 4.7.8 *Productivity*

Production rates using lightweight aggregate concrete are like the curate's egg – a mixture of good and bad.

Because of the porosity, production at the mixer is liable to be lower than with conventional mixes. If adequate pre-soaking has occurred then its effect on mixing times is removed, but the costs are moved up the line to the soaking process which can be either at the site of the mixer or at the site of the aggregate manufacturer. It would be better at the latter location as it places the cost where it belongs – with the aggregate, not the process.

Once the absorbency has been dealt with, mixing is relatively easy with no cost penalty. Delivery of the mixed concrete has penalties only when pumpable lightweight concrete is required. This creates difficulties in transportation due to spillage because of the very high workability necessary, i.e. 500–700 mm on the flow table (200 mm + slump)!

Pumping for general purposes is easy, provided that the mix is consistent and controlled for the right workability range. Again there is a cost as the mix must be carefully supervised to ensure that any adjustments are properly carried out. Very long or high-rise pipelines require careful mix design but provided that control is adequate, then lightweight concrete can be pumped quickly and economically.

Five inch pipelines are desirable for normal use, but essential for extremes, although a 2.5 in line can be used effectively with careful mix design and high-pressure pumps. This is an economic solution for small inaccessible pours and infill work.

When lightweight aggregate concrete is being used on profiled metal formwork, full advantage can be taken of the low tendency to shrinkage cracking which, together with the strain distribution of the metal deck and structural steel grid, enables large pour sizes to be placed, limited only by the resources which can be brought to bear and by economic restraints. Obviously, for these large pours, preparation and planning is important.

Compaction is also easy, and the labour necessary to place and compact lightweight aggregate concrete is less than for normal weight concrete, the material itself being lighter. Once in place it presents no obstacles to finishing, except for the tendency of the aggregate to rise, requiring power floating for smooth finishes. General spreading and levelling required far less effort than normal weight concrete, being more akin to super-plasticised flowing concrete with most of the advantages of that material. Segregation is not usual and the minimum of vibration is required. Indeed, for pumped lightweight concrete, tests have shown it to be self-compacting.

Once the initial difficulties caused by absorption have been overcome, the labour content in the use of lightweight aggregate concrete is less than for normal weight concrete which, together with the larger pour sizes possible, goes some way to reducing the penalties in dealing with absorption.

## 4.8 Economics

There have been a number of studies investigating the effects on the cost of construction from using lightweight aggregate concrete. In October 1987 the Institution of Structural Engineers with the Concrete Society produced the *Guide to the Structural use of Lightweight Aggregate Concrete*, which showed that the advantages in weight saving more than compensated for the higher initial material cost. In 1980, a study in the United States analysed a multi-span precast segmental post-tensioned box girder bridge and showed that savings achieved by using lightweight instead of normal weight concrete were 18% in the superstructure and 6% in the substructure. In 1986, a similar study in Great Britain by the TRRL and the Concrete Society showed that a 1.5% saving could be achieved in a relatively small single-span *in-situ* reinforced concrete bridge. It was expected that greater savings would be achievable on longer, multi-spanned bridges.

In 1984, a working party of the Concrete Society Lightweight Concrete Committee made a study of the economics of construction of a multistorey reinforced concrete-framed building. Their conclusions were, that from the most conservative approach there was no cost penalty in constructing in lightweight aggregate concrete, and that if 'spin off' advantages were taken into account then there were positive economic gains to be had in multi-storey structures.

These studies contradict the commonly held view that lightweight concrete is an expensive building material suitable only for exotic structures and profiled metal composite construction. The benefits to the design of the reduced dead load outweigh the higher initial costs, particularly if the structure is designed to take full advantage of the reduced dead loads – increasing spans to improve lettable space and thinning floor and beam sections to reduce heights. Figure 4.8 shows thin, precast floor elements being lifted into position.

The studies examined only the direct costs of construction and took no account of the effect on the value of the structure. The economic audit of a structure should not only examine the costs to build, but should also examine the running and maintenance costs, durability and longevity. It should, in addition, examine the ability of the structure to sustain its integrity when subject to unforeseen accidental damage. Fire is the most common of such occurrences.

For the building owner and operator, there are other benefits which add value to the structure and reduce its operating costs. Cladding and

**Figure 4.8** On-site precast ceiling units for Lloyds headquarters building, Chatham.

floors made from lightweight concrete have an improved insulating value, reducing heating costs. The resistance to fire damage of structural elements is well established. There have been examples in the UK, the United States, Japan and Australia where the use of lightweight concrete has significantly reduced the effect of fires in structures, in some cases making the difference between an inconvenience and a disaster.

Developers and owners are not so influenced by this property, as for them fires are rare and costs are borne by the insurers. Insurance companies, however, should recognise that by resisting fire damage, lightweight concrete structures reduce remedial costs and, probably more important, reduce the 'down time' of the building while repairs are carried out.

If insurance companies included these facts in their risk assessment and adjusted their rates accordingly, there would be a further benefit to the owner/operator of the lightweight aggregate concrete structure.

If the ecological benefits of using the waste materials which source many lightweight aggregates were to be recognised, and if financial encouragement were to be given for its use, then the economic sum would further favour lightweight aggregate concrete.

## 4.9 Conclusions

In general construction, lightweight aggregate concrete occurs in a relatively small proportion of reinforced concrete frames and bridges, and, where the designer was familiar with its advantages, in a number of important stadia. Its most common use is in the construction of concrete floors on profiled metal formwork.

It is more expensive per cubic metre than normal weight aggregate concrete, but if full advantage is taken of its properties, savings can be made by the design which more than offset the material costs. Where dead weight/live load ratios are high, then the balance swings in favour of lightweight aggregate concrete.

There are misconceptions and unfounded concerns about its use which can only be dispelled by experience and knowledge of its characteristics. The most significant property after lightness is its absorbency. Once the procedures have been altered to take account of this, then lightweight aggregate concrete's other positive properties can be appreciated. It is consistent and predictable, capable of high strength/weight ratios. Its lightness allows savings in handling, plant and labour. It is less prone to shrinkage cracking, so allowing larger pour sizes.

Formed finishes can be of the highest quality and power-floated finishes are also good. There is a tendency for the aggregate to float in other unformed finishes.

Curing is, as for any concrete, important, but the absorbed water enables a degree of self-curing which reduces the effects of insufficient curing.

Pumping, providing absorption is dealt with, is successful, and handling and placing are trouble free.

Most problems in practice are the result of unfamiliarity with the material. As with all construction experience, trained teams following good practice reap the reward of success.

The concrete producer and constructor need to understand the characteristics which require them to modify their methods to suit, but once these characteristics are understood and the methods adjusted, then lightweight aggregate concrete becomes a reliable, consistent material capable of high strengths, which is easy to transport, place and compact. Its weight reduces plant and labour costs. It is easy to cut and drill. It can be placed in larger pour sizes.

In all, the use of lightweight aggregate concrete should be priced on consideration of the known requirements, with no need to load the price for unpredictable potential costs.

## Acknowledgements

The author would like to thank all those who have helped him with the production of this chapter, especially Tony Elsworth of the Pochin Group, Mike Murphy of RMC Technical Services, George Lory of Starmin (England) Ltd and Mike Staff of Pozzolanic Lytag.

## Appendix

### Procedure for receiving and controlling Lytag for site pumping

The readymixed concrete will be delivered with a target workability of between 550 and 600 mm as determined for each individual contract. This will ensure that no individual load exceeds the maximum flow, but will result in a number of loads being below a pumpable workability on arrival at site.

The workability will be measured by the flow table and, dependent on the result, water will be added under the strict supervision of a suitably qualified person, in accordance with the following:

| Flow table reading | Action |
| --- | --- |
| Below 500 mm | Reject |
| 500 mm | Add 25 litres of water per cubic metre of concrete, re-mix and test again |
| 525 mm | Add 15 litres of water per cubic metre of concrete, re-mix and test again |

| 550 mm | Add 10 litres of water per cubic metre of concrete, re-mix and test again |
| 575 mm | Add 5 litres of water per cubic metre of concrete, re-mix and test again |
| 600–650 mm | Accept |
| Above 650 mm | Reject |

Where the quantity of water shown above does not provide the required workability, the process should be repeated.

The above will be agreed with both the readymixed concrete supplier and Messrs Lytag Limited and in no way invalidates the terms and conditions of the readymixed concrete supplier.

These procedures are subject to review should it become apparent that the above procedure requires modification, i.e. additional cement, sand, pump aid or levels of water adjustment.

# 5 Lightweight concrete in buildings

D. LAZARUS

## 5.1 Introduction

### 5.1.1 General

This chapter looks at the use of structural lightweight aggregate concrete in buildings. It examines the different structural applications primarily in the United Kingdom but with some reference to construction overseas. It also considers the factors affecting construction costs and comparative economics of buildings when lightweight aggregate concrete is used. It does not cover the use of very lightweight mixes of low strength in, for example, insulating screeds and other non-structural applications.

### 5.1.2 Historical perspective

Lightweight aggregates were first used in buildings over 2000 years ago when the Romans used natural pumice aggregate in concrete for such notable edifices as the Pantheon. In the United States processed lightweight aggregates were first manufactured before the end of the First World War, around 20 years earlier than to the east of the Atlantic. Lightweight aggregate concrete has been used in North America in a substantial number of both bridges (estimated at over 150 in the US and Canada [1]) and high-rise buildings.

Lightweight aggregate concrete has been used in Japan for nearly 30 years, in buildings and in civil engineering structures. In Australia, similarly, use of the material goes back over 20 years. It appears here that its use is determined partly by the availability locally of naturally occurring lightweight aggregates ('scoria').

More widespread use in Britain, firstly involving clinker and subsequently both processed by-products, such as blastfurnace slag, and manufactured aggregates, dates from around the beginning of this century.

Since then a variety of both categories has been produced and such aggregates are now in more common use. While use of these materials both removes the need to dispose of industrial by-products and decreases the requirement for naturally occurring, and therefore ultimately limited, materials their acceptance in recent years has been somewhat qualified. Unfamiliarity with the rules governing design – which are perhaps wrongly perceived both as differing widely from those for the design of structural concrete using normal weight aggregates and as being unnecessarily restrictive – and a lack of accurate and detailed comparative cost information tend to inhibit the use of lightweight aggregates. An attempt to overcome the former was made in 1991 with the production of the 'Designer's Fact Sheets' for the design of lightweight aggregate concrete to BS 8110 [2] by the Concrete Society's Lightweight Concrete Committee [3]. These were in fact the successors to information first produced nearly 40 years earlier: *An Introduction to Lightweight Concrete* first published by the Cement & Concrete Association in 1953 (revised in 1954 and 1960), and the Concrete Society Data Sheet 1980, 'Lightweight Aggregate for Structural Concrete'.

At the present time, use of lightweight aggregate concrete in the United Kingdom is more restricted than in both North America and Continental Europe and, in general, it has perhaps been used less adventurously than overseas where more tall buildings incorporating the material have been constructed. Nevertheless, a substantial number of examples of its use, some of which go back more than 25 years, can be seen both in London (e.g. the Chinese Embassy, Guy's Hospital) and elsewhere (e.g. Students' Residences at the University of East Anglia, Redbridge Magistrates Court).

## 5.2 Factors in the selection of lightweight aggregate concrete

### 5.2.1 Introduction

There are many factors which influence the selection of lightweight aggregate concrete. These are discussed here and elsewhere in the book. It is worth looking at some of these here in more detail before moving on to consider more specific applications of the material in buildings.

It is important to recognise that substituting lightweight aggregate concrete for normal weight concrete in the same structural configuration will not necessarily result in the optimum use of the material, and may well not constitute good design. In the brief case studies presented later in this chapter it will be seen that for many concrete-framed buildings only particular elements are constructed in lightweight aggregate concrete; this is in order to take advantage of particular properties of the material in specific areas, perhaps to achieve a required finish or structural configuration.

### 5.2.2 Durability

The previous section referred briefly to the existence in this country of a number of buildings constructed wholly or partially in lightweight aggregate concrete more than 25 years ago. A survey of 25 of these (plus seven bridges and eight 'special structures') containing exposed concrete in structural elements and all built earlier than 1977 was carried out recently by Mays and Barnes [4]. A visual inspection of the buildings was undertaken and for five of them a more detailed examination, including core testing and covermeter surveys, was carried out.

There has been concern among engineers that lightweight aggregate concrete would have a higher permeability than normal weight concrete due to the aggregate porosity. BS 8110 appears to formalise this by the requirement for 10 mm additional cover to the reinforcement for all conditions of exposure other than 'mild'. This would seem to ignore both the beneficial effect on the rate of carbonation of the higher cement content, and perhaps in particular the resulting lower water/cement ratio which is required for an equivalent strength lightweight aggregate concrete mix and the coating to the aggregate offered by the cement paste.

Mays and Barnes concluded on the basis of their observations and a review of previous data, including research, that there was no evidence to support the supposed reduction in durability of lightweight aggregate concrete provided that adequate attention was paid to specification and workmanship. They warned of a possible increase in sensitivity to poor workmanship when compared with normal weight concrete. Earlier structures tended to use lightweight coarse and fine aggregates; some research has suggested that carbonation resistance is better for a mix with natural fines, so it could also be argued that more recent structures will show an improved performance. This may be due to the requirement for more water, and hence a higher water/cement ratio where lightweight fines are used.

The Japanese and American experience, based on investigations of structures built up to 20 years previously, had also confirmed the comparable durability of concrete made with lightweight aggregates for mixes with a low water/cement ratio and thorough compaction, i.e. similar considerations applied to those required for the specification of normal weight mixes.

### 5.2.3  Fire

The reasons for the superior performance of lightweight aggregates when compared with the majority of normal weight aggregates are well known, and are discussed both in the chapter by Clarke and in standard references. This is also the area where a benefit is specifically recognised in BS 8110 by allowing lower cover to reinforcement than for normal weight concrete for the same period of fire resistance. Costs of reinstatement are reduced accordingly for lightweight aggregate concrete compared with the performance of normal weight concrete in an equivalent fire. This has been illustrated in the performance of lightweight structures during a number of fires that have occurred, although data on this topic remains limited. This comparison is made for *in-situ* structures; it appears that in composite construction (see section 5.3.2) while the slabs do not collapse at high temperatures the deck separates from the concrete and replacement of the affected areas is required.

### 5.2.4 High-strength concrete

At present the term 'high-strength' when applied to lightweight aggregate concrete implies values which may be up to $100\,N/mm^2$, whereas for normal weight concrete values of around $140\,N/mm^2$ and higher are covered by this category. Although the cement-aggregate bond is probably better for the former, the strength is limited by the high water demand and ultimately by the crushing strength of the aggregates themselves.

In general the experience for lightweight aggregate concrete has been that the density of the material increases with an increase in strength, as the cement content rises, so that while a $20\,N/mm^2$ mix might be achieved for a density of $1200\,kg/m^3$, this value would increase to 1800 for an $80\,N/mm^2$ mix. The data is specific to a particular aggregate combination; the results would not replicate with the use of a different lightweight aggregate. In the United Kingdom strengths of $60-70\,N/mm^2$ have been achieved using a Lytag mix at $1850\,kg/m^3$, and current research is aiming at producing higher strengths. To date results up to a maximum of $110\,N/mm^2$ have been achieved using a proportion of silica fume.

In Germany research into 'optimised' concrete is also being carried out, aimed at increasing strength in the lower density range. To achieve this the water/cement ratio is reduced and superplasticisers used, and cement replacement with pozzolana is adopted to improve the matrix. A strength of $50\,N/mm^2$ has been achieved for a $1350\,kg/m^3$ oven-dry density, with a strength range overall of around $20-75\,N/mm^2$ for a density range of $1000-1500\,kg/m^3$. Results obtained from laboratory tests on optimised concretes using lightweight aggregates with different bulk densities from 300 to $800\,kg/m^3$ are shown in Figure 5.1 [5]. Although it is harder to achieve higher strengths when using lightweight fines, the research is also investigating this area.

The outcome of the European research currently in progress will obviously be an important factor in determining the future use of lightweight aggregate concrete in buildings, perhaps in particular as an alternative to steel-framed construction but using similar composite slabs. At this point in time, however, it has to be recognised that these 'high-strength' mixes are not accepted in some of the design codes, including the new Eurocode; this is discussed in the chapter by Clarke, and there is no reason to believe this is likely to change in the near future. The

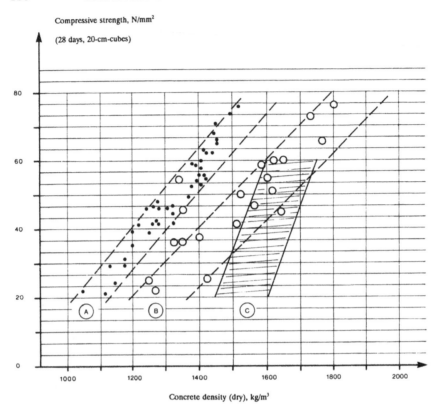

**Figure 5.1** Compressive strength and density of lightweight aggregate concretes. (After Spitzner, ERMCO Congress, June 1989.) Each ○ represents one building, ● stands for laboratory results. A, usual range in North America; B, usual range in Europe and Japan; C, improved mix design (ADV).

current $60-70\,\text{N/mm}^2$ range may thus remain at the upper end of the recognised values, notwithstanding the ongoing research.

### 5.2.5 Placing lightweight aggregate concrete

Initial concerns about the suitability of lightweight aggregate concrete for pumping have in general been overcome with the development of suitable mixes using natural sand and specially developed admixtures. Care is required to monitor the required mix proportions and to obtain a relatively uniform density. At the present time these mixes tend to be wet ('sloppycrete'); it is possible that changes in pumping technology and improved control of water absorption of the aggregate could lead to satisfactory pumping of drier (perhaps 75 mm slump) mixes.

Guidelines have been produced by lightweight aggregate manufacturers for pumping concrete to the heights required in modern medium- and high-rise buildings. The majority of the problems encountered relate to the absorbency of lightweight aggregates and the need for frequent checks on its moisture content and density. The time taken for thorough mixing of the concrete constituents is greater than for normal weight concrete, and stiffening of the mix in transit is inevitable to some degree.

It is essential to agree special procedures, such as addition of water on site, for any project before concreting starts and to ensure that all parties understand the use and properties of the material.

Water absorption of the aggregate may be controlled by pre-soaking or by the use of appropriate admixtures. The former is used widely in America, with three methods being available:

1. vacuum soaking to produce a virtually saturated aggregate
2. immersion in a tank of water with continual agitation for a specified time
3. continual sprinkling for three to seven days.

Of these the first is the most efficient, but all are expensive and have not been adopted in the United Kingdom. The exception to this is Canary Wharf, which to date holds the record for the highest use of pumped lightweight aggregate concrete in the United Kingdom, where vacuum soaking was used; this is particularly advisable when pumping on high-rise (>40 m) buildings. Further information is given in chapter 4 on 'Construction'.

### 5.2.6 Slipforming

There is no inherent reason why slipforming should be any less successful for lightweight concrete than for the normal weight equivalent in similar circumstances; the requirements for a suitably cohesive mixture and the correct setting-time balance are the same in both cases, and there are no particular problems noted which apply to one material rather than the other, although because of long travel distances to the top of the slip it is important to control water absorption by the aggregates. Problems with slipforming are also more likely to occur where flaky aggregates are used and/or with gap-graded aggregates, neither of which should legislate against the use of lightweight aggregates. There are, however, relatively few instances of this technique being used, although it has been utilised in a few high-rise buildings in the United States, for example the Raymond Hilliard Centre in Chicago (see 'Brief case studies' on p. 117); in many instances it appears that for such buildings the walls are constructed in normal weight concrete, where slipforming is more common.

Two examples which have been noted in this country relate to the construction of sugar silos, and both go back over 20 years. The earlier of the two is the combined silo and elevator tower built at Spalding in 1965. The concrete for the elevator tower was an Aglite–sand mix with a water/cement ratio of 0.67, a 50 mm slump and a 28-day strength of 20 N/mm². The high water/cement ratio was specified to take account of water absorption of the lightweight aggregate. The walls were 280 mm thick and the average climbing rate was 350 mm per hour. The second project, the conveyor tower at the Wissington sugar silo, represented the first use of sliding shutters with Lytag concrete. Again, natural sand was used in the mix with a design strength of 25 N/mm² at 28 days. The Lytag was pre-soaked to control the water demand. Almost 350 m³ of concrete were placed in the walls, which reached a height of over 60 m, and the rate of lift was recorded as being 'as normal for dense concrete' [6].

### 5.2.7  Finishes

Sufficient experience has now been acquired within the industry to pump concrete satisfactorily subject to the caveats described in section 5.2.5. The flowing mix may produce a virtually self-levelling finish, and in a few instances it has even been found that little if any treatment is required for the installation of a raised floor.

Slabs may be power floated or trowelled as required, although opinions differ as to the possibility of power floating lightweight aggregate concrete. Both consultants and contractors have been recorded as saying that it cannot be done. Again this bias seems to have arisen at least to some extent from experiences with high-workability concrete, where power floating is certainly much harder because the lightweight aggregate floats to the surface of the denser matrix. In the higher flow mixes adopted for easier pumping the floating process tends to drag the aggregate, thereby creating an uneven surface.

It should, however, be emphasised that while sloppycrete is widely regarded as the norm for lightweight aggregate concrete, arising from the widespread use of pumped mixes for composite construction, slumps of the order of 75 mm are perfectly feasible and will commonly be specified for more conventional *in-situ* and precast use.

Good formed finishes are perfectly achievable, due in part to the high cement and fines content necessary to achieve the specified strength and durability. The high quality of finishes which can be obtained is exploited in those buildings where exposed concrete surfaces are specified. Examples are quoted in 'Brief case studies' on p. 116.

### 5.2.8  Finishing

Lightweight aggregate concrete may be rendered, plastered or painted; in this respect there is little difference from normal weight concrete. This is referred to again subsequently in section 5.3 in relation to the use of precast lightweight aggregate concrete panels for domestic and light-weight industrial construction.

## 5.3  Applications of lightweight aggregate concrete

### 5.3.1  In-situ concrete structures

*(a) General.*  The structural benefits of lightweight aggregate concrete are probably most significant for medium to long-span structures, where self-weight is dominant and where the ground conditions make it particularly important to keep the weight of the building down. In addition, the benefits relating to speed and ease of construction discussed later in the chapter are realised. For high-rise buildings these factors are especially applicable; the form and construction of the floor slabs is possibly the most critical aspect of the construction in relation to programme and there are several instances where lightweight aggregate concrete is used only for all or part of the horizontal structure. The thermal properties of the material have dictated its use for external walls and cladding where an exposed finish is required internally and externally, i.e. without the use of added insulation.

Despite the acceptance of lightweight aggregate concrete for a variety of structural uses, the number of instances where it is used for *in-situ* construction remains limited, perhaps particularly in the United Kingdom. This is especially true when considering buildings where the complete frame, as opposed to, say, floors or supporting beams, is constructed using reinforced lightweight aggregate concrete.

The British Cement Association published a survey in 1990 of 40 office buildings with long-span concrete floors [7]. This was commissioned by the Reinforced Concrete Campaign Group to verify the competitiveness of long-span concrete floors. Of the 40 buildings selected, 33 were of *in-situ* construction (reinforced or prestressed) and the remaining seven utilised precast or composite precast slabs. Of these, only three of the *in-situ* reinforced concrete buildings are noted as being of lightweight concrete.

Instances of post-tensioned lightweight concrete slabs are even less common: in a recent investigation only three examples were found, one of which, with spans of 12.6 m, is included in the BCA publication referred to above. The advantages of reduction in weight, and hence in prestress requirements, may see this form of construction becoming more widely considered. At the present time it is believed that the need to use additional steel to offset increased creep and shrinkage losses, combined with lack of experience in the technique for both design and construction, tend to legislate against its use.

One such example was, however, constructed in Cheltenham in 1983/84 to form an extension to Mercantile and General's offices. The consulting engineers for the project were Clarke, Nicholls and Marcel. The use of post-tensioned flat slabs for the five-storey extension enabled the requirement for large spans to be met while matching existing storey heights. Bay sizes of 11 m by 9 m were used, with a 440 mm slab (one-directional stressing) at first floor and a 375 mm slab (two-directional stressing) at the upper levels. Lytag concrete was used, with each floor cast in one continuous operation; pours of up to 300 m$^3$ were placed in one day, using two pumps and two readymix plants supplying a target of 40 m$^3$ per hour.

A further example of post-tensioned structural lightweight aggregate concrete may also be seen in London, in the roof to the vehicle maintenance depot for the Royal Borough of Kensington and Chelsea. This is illustrated in Figure 5.2.

*(b) Brief case studies.* The 40 buildings investigated by the BCA were generally of relatively recent construction. The first framed building constructed in Britain using lightweight aggregate concrete, a three-storey office block in Brentford, was constructed in 1958. This is some 30 years behind the United States where the Park Plaza Hotel in St Louis and the South Western Bell Telephone Company in Kansas City were both built in 1928 using this material.

Thereafter a small number of buildings were produced during the 1960s and 1970s and this limited output has continued. Again this does not

**Figure 5.2** Kensington & Chelsea Maintenance Depot. (Arup Associates, Architects, Engineers & Quantity Surveyors.)

**Figure 5.3** Roxburgh County Offices: external view. (Ove Arup & Partners.)

mirror the American experience where lightweight aggregate concrete was used increasingly widely; by the mid-1950s it was being used in high-rise buildings, with the 42-storey Prudential Building (now the Prudential Plaza Building) in Chicago built in 1955. This is notable for several reasons: when planned it was the fifth largest office building in the world, and on completion it was the tallest building in Chicago. It was also the first building constructed on air rights of the Illinois Central railway yards [8].

Details of a number of significant buildings where lightweight aggregate concrete was used are given below.

1. For Roxburgh County Offices, built in 1966/67, lightweight aggregate concrete is used for all elements of structure other than the foundations and basement. This was the first complete structure built using lightweight aggregate in Scotland (see Figure 5.3). Graded Lytag (sintered PFA) aggregate and natural sand were used. Full-height cladding panels and spandrel beams, which are board-marked and ribbed on

**Figure 5.4** Roxburgh County Offices: structural mullion detail. (Ove Arup & Partners.)

both internal and external surfaces, are also of lightweight aggregate concrete. The structure comprises coffered floor slabs supported on slender perimeter mullions which were precast; the building is square in plan with an internal courtyard at the centre.

The use of lightweight aggregate concrete was specified here to take advantage of three particular properties of the material:

(a) The reduction in dead load due to the floor slabs enabled very slender supporting mullions to be used (see Figure 5.4).
(b) The improved fire resistance again allowed the mullions to be particularly slender while complying with the requirements of the Building Regulations, as a lower cover could be used than that required for normal weight concrete.
(c) The improved thermal insulation of the spandrel beams and cladding panels permitted the required exposed concrete finish to be adopted both internally and externally, and the absence of applied finishes also enabled a short construction period to be achieved.

This is a good illustration of the application of particular benefits of the material, in this instance resulting in its utilisation throughout the structure rather than for specific elements only.

2. Marina City in Chicago was built in the early 1960s on the banks of the Chicago river. Its twin cylindrical towers provide 60 storeys of apartment blocks, reaching some 179 m high. Here the floor slabs were cast using lightweight aggregate concrete in an interesting early

use of the material in the United States, apparently to enable a reduction to be achieved in the length of the vertical elements. The aggregate used was an expanded shale, 'Materialite'. The thin slabs are expressed on the outside of the building, in particular on the first 20 storeys which accommodate the parking and service areas [8, 9].

From the 21st storey upwards the apartments fan out from the central cylindrical core to form 'petals' around the perimeter of the building (see Figure 5.5).

3. The Raymond Hilliard Centre in Chicago was built in the mid-1960s. Designed by the same architect, Bertrand Goldberg, as Marina City the form of the building has distinct similarities and uses the material in the same adventurous manner. Both buildings illustrate clearly the sculpted forms which can be achieved with concrete, and provide a strong contrast with the rectilinear steel-framed towers which have proliferated in Chicago. Here all columns, loadbearing walls and floor slabs are constructed from lightweight aggregate concrete, with 'Materialite' coarse and fine aggregates used together with natural sand. All external surfaces are board marked and the spectacularly curving walls were slipformed [8, 9].

4. One of the most interesting examples from the 1970s in the United Kingdom is Guy's Hospital, London, for which R. Travers Morgan & Partners were the consulting engineers. This is perhaps a classic case of high-rise construction dictated by limited site availability, with lightweight aggregate concrete used to reduce foundation loads not only in structural elements but also in the external walls.

Two tower blocks, the 'User Tower' and the 'Communication Tower' were built; these were, respectively, 122 and 145 m above ground floor, each with a lower ground floor storey and single basement. The benefits of reducing the dead load were also utilised in the method of construction, so that at the upper levels the amount of prefabrication in the floors was maximised. The first five levels of superstructure were constructed using solid lightweight aggregate slabs; above this level a special ribbed floor construction was developed (see Figure 5.6) [10]. Lightweight aggregate concrete with a fluted profile, achieved with the use of special steel formwork, was also used for the external walls of the higher tower which were cast *in-situ*. In certain areas advantage was taken of the properties of lightweight aggregates to provide the necessary two hours' fire resistance without a requirement for screed. A dramatic use of lightweight aggregate concrete was in the lecture theatre on the 29th floor. The material was selected here in preference to other structural options considered to construct a series of raking beams cantilevering into space 113 m above ground and supporting concrete cladding.

The structure was designed to CP 114 [11]. The mix specified was

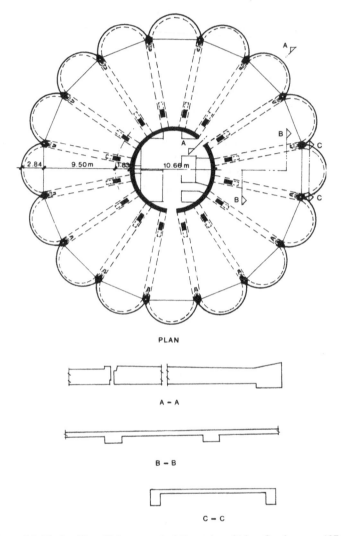

**Figure 5.5** Marina City, Chicago: typical floor plan. (After Cembureau, 1974.)

to have a 28-day strength of 4500 lbf/in$^2$ (31.3 N/mm$^2$), and both Lytag coarse and fine aggregate were used, with 390 kg/m$^3$ of cement.

5. The NLA Tower in Croydon was also built over 20 years ago, utilising 250 mm solid flat slabs spanning just over 8 m. Ove Arup & Partners were the consulting engineers. In this building only the slabs are of lightweight concrete: all supporting structure, including the floor beams, is of normal weight concrete.

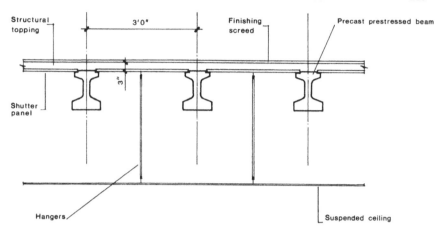

**Figure 5.6** Guy's Hospital, London: details of ribbed floor construction.(After *The Structural Engineer*, 1971.)

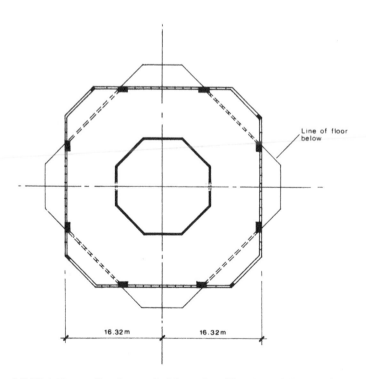

**Figure 5.7** NLA Tower, Croydon: typical floor plan. (Ove Arup & Partners.)

Here the alternative floor systems investigated were narrowed down to two choices dictated by the unusual plan form of the building (see Figure 5.7). The two options were:

(a) a 10 in (250 mm) solid lightweight slab or

(b) a 10 in coffered slab in normal weight concrete.

Both options were included for pricing by the tenderers and five of the seven tenders received, including the successful one, showed a cost saving for the lightweight option.

Again a Lytag coarse and fines mix was specified, with a 28-day cube strength of 4200 lbf/in$^2$ (29 N/mm$^2$). Cement content was 297 kg/m$^3$ and the concrete was site-mixed.

6. In the mid-1980s perhaps the most significant example of an *in-situ* reinforced lightweight concrete building constructed in London was the new Chinese Embassy in Portland Place, designed by Jan Brobowski & Partners. This building has seven storeys above ground plus a part-mezzanine, together with a full basement and a partial second basement level. Lightweight aggregate concrete is used for the beams and suspended slabs at all levels and for the sloping walls on the upper storeys; a typical arrangement is shown in Figure 5.8. The slabs are 175 and 220 mm thick, spanning between 4.4 m and 6.75 m. Here a Lytag mix was again used but in this instance only for the coarse aggregate. With the advent of pumped concrete it was found that natural fines were required to produce a readily pumpable mix. A characteristic strength at 28 days of 43.5 N/mm$^2$ was achieved, based on 105 cube test results (specified 30 N/mm$^2$), with a cement content of 396 kg/m$^3$ (277 kg OPC, 119 kg PFA).

7. One Shell Plaza Tower in Houston, built in 1967, is a famous example of the use of lightweight concrete in an *in-situ* structure; in this instance all structural elements incorporate lightweight aggregates. At approximately 220 m high this was the tallest lightweight concrete building in the world when it was completed.

Typical spans are 11 m, and the covered floor area is 130 000 m$^2$. The concrete used had a density of 1840 kg/m$^3$ and a 28-day cylinder strength of 42 N/mm$^2$. This is an example of double-tube construction, with closely spaced columns in both the external wall and the central core.

Perhaps the most significant feature of this building is the achievement of a 52-storey structure on a raft foundation replacing the 35-storey building in normal weight concrete schemed on a similar foundation in the same depth of excavation [9].

8. Australia Square in Sydney was built at a similar time, and at 184 m high is almost as tall. It comprises a 56-storey tower of 42.5 m diameter above a five-storey base.

Above the seventh floor the entire frame is built of lightweight

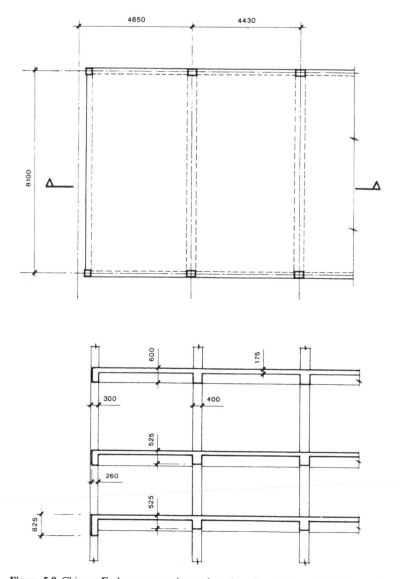

**Figure 5.8** Chinese Embassy: part plan and section. (Jan Brobowski & Partners.)

aggregate concrete using 'Litecrete', an expanded shale. The floor slabs, providing two hours' fire resistance, are thinner than would be permitted under the Australian code using normal weight concrete. A saving in building weight of 38% was calculated for the lightweight concrete option.

Lightweight concrete was also used for precast permanent formwork for spandrel beams and the tapering perimeter columns [9].

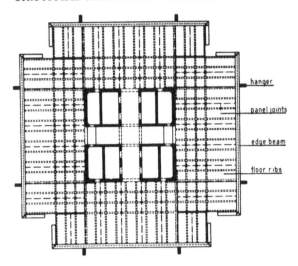

**Figure 5.9** Standard Bank, Johannesburg: typical floor plan. (Ove Arup & Partners.)

Two buildings built in the late 1960s/early 1970s in very different locations used lightweight aggregate concrete for the same reasons in a similar structural configuration. In both cases lightweight concrete is used for part of the structure only. The finished buildings, however, have very different appearances.

9. The BMW administrative building in Munich is effectively of clover-leaf form on plan. It has a central core and four slender central suspension columns which are anchored to a supporting structure cantilevering from the top of the core. The building is 100 m high with 17 floors plus a services floor suspended. Only the suspended floors are constructed in lightweight aggregate concrete, specifically to reduce the dead load of the building. Ribbed slabs in concrete grade LB300 (30 N/mm$^2$ lightweight concrete) result in deck weights which are no heavier than normal weight concrete on metal deck; they are designed as fixed at the connection to the hangers and pinned at the external column line bearing on neoprene pads over nibs. Liapor (expanded shale) coarse and fine aggregates were used, together with natural sand [9, 12].

10. The Standard Bank building in Johannesburg is 139 m high and contains 30 storeys. Its plan form, shown in Figure 5.9, appears rather more conventional by comparison with the BMW building. The similarity lies in the unusual suspended structural form used in both buildings and the use of lightweight aggregate concrete in the floor slabs only in order to reduce dead load. The floor slabs

**Figure 5.10** Standard Bank, Johannesburg: floor construction. (Ove Arup & Partners.)

in this building were constructed using precast double-T units which were steam-cured and lifted one day after casting to achieve a rapid erection time.

The units are 10 m long and up to 3.16 m wide; the slab is 75 mm thick between ribs at 1.58 m centres. These are shown in Figure 5.10. Aglite, an expanded clay, was used for the coarse (20/10 mm blend) aggregate with natural sand, giving a relatively dense mix (1950 kg/m$^3$ at 28 days). The hanging structure allowed the creation of an open area at ground level around the central core, without the need for deep transfer structures at the level of the lowest suspended slab [9, 13].

11. Lake Point Tower in Chicago was constructed in the late 1960s. It is possibly the tallest apartment building in the world. Standing 70

storeys and some 200 m high, the cloverleaf plan is symmetrical about a triangular core, and the elegant curves of the glazed tower appear to owe something to the influence of Mies van der Rohe.

Lightweight aggregate concrete is used in the apartment floor slabs, which are 200 mm thick, and also in the deeper waffle slabs which are designed to carry heavier loads in the four-storey base structure housing parking and service facilities below the tower. The aggregate used was 'Materialite' [8, 9].

12. The Commercial Centre Tower at Kobe, Japan, was constructed in the late 1960s. It stands 109 m high above ground. All floors from the 3rd to the 26th, including the roof, were constructed using lightweight concrete. The aggregate used was 'Lionite', an expanded shale [9].

13. Central Square office building in Sydney was built in 1971/72. It is a 26-storey building with a central core, internal columns and structural mullions. The structure is innovative in two details, primarily the use of steel trusses as reinforcement for secondary beams and also the adoption of a lightweight aggregate concrete core for the mullions, poured *in-situ* using precast concrete 'formwork'.

Lightweight aggregate concrete, in this case using 'Litecrete', an expanded shale, was widely utilised in the structure. Lightweight concrete elements include the core walls and columns above the 13th level, the mullion core noted earlier, the primary beams and the floor slabs, which were poured integrally with the secondary beams.

14. At around the same time as this some smaller scale buildings were being constructed using lightweight aggregate concrete in the United Kingdom.

Pimlico Secondary School is built entirely of lightweight aggregate concrete. Lytag coarse and fine aggregate was specified with no natural fines. The concrete had a density of 1680 kg/m$^3$ and is used for floor slabs (with significant cantilevers), beams, portal frames and loadbearing walls. The concrete is exposed both internally and externally in the building, with a board-mark finish on the wall panels on the outer face and a smooth finish on the inside. Lightweight concrete was selected here to suit particular performance requirements, specifically to provide the required level of insulation to enable exposed concrete to the used, and the savings which resulted more than offset the cost premium on the lightweight aggregate itself. It is interesting to note that the same mix was used for a large external water tank [9].

15. The Sheraton Park Tower in Knightsbridge is 15 storeys high above the podium at first floor. The tower is circular on plan with a diameter of approximately 37 m. Lightweight aggregate concrete with Lytag coarse aggregate and natural fines was specified for all elements above the podium transfer structure: 175 mm solid slabs, radial walls

and the central core. This was in order to reduce the foundation loads and minimise the number of columns below the podium.

16. Lightweight aggregate concrete was used at the British Library (Phase 1A) for the long-span *in-situ* beams and columns which make up the entrance hall framework, and also for the precast roof slabs over both the entrance hall and the Reading Rooms. Lytag coarse and fine aggregate was used in the mix.

For the *in-situ* framework a 40 N/mm² mix was specified, with a minimum cement content of 510 kg/m³, a 75 mm slump and the use of a plasticiser (w/c 0.44). The 28-day cube strengths were generally in excess of 50 N/mm², with most 7-day strengths well over 40 N/mm².

For the precast slabs the minimum cement content was 450 kg/m³, total water 295 kg and a 50 mm slump. The mean of 38 No. 7-day tests was 47.5 N/mm² and that of 28 No. 28-day tests was 61.3 N/mm².

The arrangement of the entrance hall pitched roof structure is a special feature; the beams span 15.6 m and the framework creates a large clear space 18 m high. The structural solution was adopted in preference to a steel frame for its enhanced durability and fire-resistance properties.

17. The Torre Picasso building in Madrid was built in 1988/89. This stands 150 m high and contains 3500 m³ of lightweight aggregate concrete in the floor slabs. The concrete used was 45 N/mm² with a dry density of 1650 kg/m³ and was placed by pumping.

The above demonstrates a number of applications of lightweight aggregate concrete worldwide. It is noticeable that the extent to which the material is specified in any individual structure varies considerably. It may also be seen that while Lytag is used predominantly in the United Kingdom, a variety of other aggregates are used overseas. The aggregates themselves vary in type, shape, density and strength grade range. In the United Kingdom six aggregates are available, of which two are imported. These are described in the 'Designer's Fact Sheets' referred to earlier [3]. It is interesting to note that a decade earlier the Concrete Society Data Sheet 1980 'Structural Lightweight Aggregate Concrete' listed seven lightweight aggregates, of which only four appeared in the later document. The mixes also vary to a large extent, in particular with respect to the use of lightweight fine aggregate, with or without natural fines, and the resulting density of the concrete. Changes have been made as construction techniques have become more sophisticated, and further changes may be expected with the advent of higher strength concretes. These should be seen as positive developments, properly backed up with research programmes, rather than as an excuse for avoiding what may still be perceived as a 'new' material.

### 5.3.2    *Composite slabs with profiled metal decking*

While there remains some reluctance among both consulting engineers and quantity surveyors in the United Kingdom to accept the use of lightweight aggregate concrete for *in-situ* structures, very considerable quantities of the material have been utilised over the past few years in composite slabs with profiled metal decking. It is believed that around 80% of all lightweight aggregate concrete in the United Kingdom is used in this form of construction.

After a long period during which steel-framed construction was not much used due to high material costs and perceived problems of fire protection, the explosion of medium-rise office buildings in the United Kingdom in the 1980s – particularly in the south-east – saw this form of construction, long popular in North America, become predominant. The majority of these buildings used thin lightweight aggregate concrete slabs acting compositely with profiled metal decking connected with shear studs to the floor beams, and in a rapid volte-face this composite construction became the norm.

A brief survey of major developments in central London within this period indicates the extent to which both developers and designers saw this form of construction as the optimum structural solution. To the non-technical observer the initial impression in walking through the City, past the numerous sites under construction simultaneously, was that all buildings had become identical. Perhaps the best-known examples are many of the Broadgate buildings, and, most famously, the high tower of Canary Wharf. The latter, at 50 storeys and some 235 m above ground, is not only the tallest building in London (some 90 m higher than the National Westminster Tower) but represents the use of lightweight aggregate concrete at maximum height in the United Kingdom. The pours at Canary Wharf were also large, another feature of this period when pours of $300 \, \text{m}^3$ were not unusual.

Figures obtained from Boral Lytag, one of the United Kingdom's largest suppliers of lightweight aggregates, show that the volume of structural lightweight aggregate concrete supplied nationally rose annually between 1985 and 1990. During this period the annual total increased by a factor of almost 3.5 to over a quarter of a million cubic metres. In 1991 the volume fell back sharply to a figure closer to the 1986/87 figures. Similarly, industry sources showed that the quantity of decking rose very steeply between 1983 and 1990. Here, over a longer period of time, the increase was more than eight-fold, although precise figures for the quantity used in composite construction were not available.

The advantages of using lightweight rather than normal weight concrete in this form of construction may be summarised as follows:

1. Reduced dead load of wet concrete allows longer spans to be poured unpropped. This saves both labour and the cycle time for each floor.
2. The reduced dead load of the slab results in savings in the structural frame and ultimately in the foundations.
3. Improved fire performance of the material when compared with normal weight concrete. The insulation to embedded reinforcement is improved by reducing the thermal conductivity by 50%, and spalling of the concrete is very substantially reduced, although ultimately the matrix deteriorates. Overall, any loss of strength is delayed until a greater temperature is reached, but indications are that this form of construction may actually perform less well overall than a conventional *in-situ* slab using the same material.

   During the boom in office building at least two major fires have occurred in London during construction of steel-framed buildings. Some information therefore exists on the performance of composite slabs incorporating lightweight aggregate concrete in fire in actual structures rather than solely from fire tests. It appears that the deck separates from the concrete at high temperatures and that while the slab does not collapse replacement is necessary.
4. While shrinkage values are similar to those for normal weight concrete, or possibly somewhat greater, shrinkage cracking rarely occurs. This enables very large areas of slab to be placed continuously, giving a further advantage to the floor cycle time.

As the use of composite construction became more widely accepted by clients, consultants and contractors, variations on fairly conventional rectangular grids using standard UB sections were developed to suit particular requirements. At one stage there was a brief demand for longer spans (previously up to around 9 m) to create large open 'dealing floors' in the City; there have been many discussions and various alternatives proposed for minimising the overall depth of structure and services combined so as to keep storey heights down. These include the stub girder, deep beams with services penetrations and tapered beams. The concept of two-way spanning slabs has also been investigated briefly.

Recently two further developments have emerged in the field of composite construction, both of which differ from systems used previously in that they utilise a composite slab on metal decking supported on the bottom flange of the primary steel beams. For both, the use of lightweight aggregate concrete offers the same advantages as for more conventional systems.

CSC (UK) Ltd has developed the Arched Metal Deck Flooring System, for which a patent is currently pending (summer 1992) [14]. This employs similar principles to Victorian brick arch floors, and is the result of a

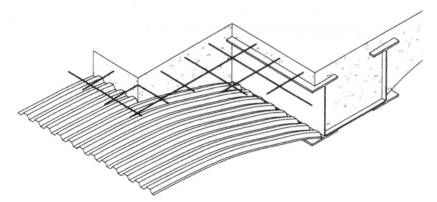

**Figure 5.11** Arched metal deck flooring system: isometric view. (CSC (UK) Ltd.)

research project undertaken by CSC for which a set of characteristics for an innovative flooring system was defined as follows:

– light rapid construction
– reduce or eliminate the requirement for structural fire protection
– increase floor spans without the requirement for construction stage propping
– allow composite action without the use of shear studs
– reduce construction depth
– provide safer fire-resistant steel structures.

The system is illustrated in Figure 5.11.

One of the main advantages of the system is the enhanced fire performance resulting from the partial protection of the supporting beams by the concrete. A full-scale fire test was carried out on a section of arched floor and compared with a similar test on an unprotected beam supporting a composite slab in its upper flange; the results for the former when measuring mid-span deflection were found to be orders of magnitude better. Test and theoretical results indicate that two hours' fire resistance can be obtained without the need for additional protection. Additional advantages are:

– improved spanning capability due to arching action (spans of up to 6 m have been tested and spans of up to 8 m appear feasible from mathematical models);
– cost savings due to elimination of both applied fire protection and shear studs;
– optimisation of overall depth of construction.

At the same time PMF have produced a new advanced profile design for metal decking. Profile 210 was also first marketed in 1992. It offers the possibility of composite floors spanning up to 6 m without the need for propping in the construction stage, and can also provide a fire resistance period of up to two hours without additional fire protection. Where lightweight concrete is used the imposed load capacity and maximum unpropped span are greater than for the case where normal weight concrete is used.

For composite construction the highly profiled sheeting is used with a British Steel 'Slimflor' beam to minimise construction depth; here the decking is supported on a continuous plate welded to the bottom flange of a compact I-section. The profile has cross-ribs which, with trough shear connectors, provide the shear-bond key with the concrete. Again there is an optimisation of the structural depth.

The first project using this system will have the decking placed in autumn 1992.

### 5.3.3   Precast units

The use of lightweight aggregate concrete in precast units, either for structural elements or as non-loadbearing cladding units, is dictated by the same benefits as those which govern its use *in-situ*.

The reduced weight by comparison with normal weight concrete not only gives an advantage in reducing the overall weight of the building but, perhaps more importantly, allows the transportation and lifting of larger units without the need for heavier plant. Handling at the works is similarly simplified. Components may be up to 20% lighter than their normal weight concrete equivalent, while providing an equivalent strength. Where units are made by an extrusion process, it has also been found that sawing costs are significantly reduced because of the lower aggregate strength.

There has been some experience of damage occurring to units during handling, apparently because lightweight aggregate concrete when freshly cured is slightly more brittle, and extra care is needed.

The thermal characteristics of lightweight aggregate concrete, as discussed previously, permit the use of exposed concrete internally and externally without the need for applied finishes and insulation. A particular example of this is the precast, prestressed cladding panel developed in the former Soviet Union more than a decade ago, for mass production to a low budget [15]. The single-layer panel provides insulation against the extreme winter temperatures using an expanded-clay coarse aggregate (keramzite) and 10–12% of entrained air. The mix has a characteristic strength of $10 \, \text{N/mm}^2$ and a density of $1200 \, \text{kg/m}^3$. The entrained air improves the durability. The panels are 240 mm thick and are manufactured in a standard size of 12 m long by 1.2 m high; they have been used in substantial quantities in both domestic and industrial buildings.

The concrete mix was developed with great care to provide the required properties and the end product is simple and cheap to produce.

A similarly innovative use of a very lightweight concrete (half the weight of normal weight concrete) is for large precast wall panels and floor slabs cast around polystyrene blocks to form rectangular voids within the panels. The concrete used is known as X-concrete; the coarse aggregate is an expanded clay, polystyrene pellets are used for the fine aggregate and the mix is air-entrained using an admixture known as AD-X [16, 17].

The use of the lightweight mix combined with void fillers allows the ready transportation and erection of large units – typically 2.5 × 6 × 0.3 m. U-values for the wall panels are in the range 0.27–0.38 W/m² K and the admixture gives water resistance to the hardened concrete. The concrete was developed in Gothenburg, Sweden, and has been used for some years in Scandinavia, primarily in housing. Despite the apparent benefits of the units, however, attempts to introduce both this and similar systems into the United Kingdom in recent years appear to have foundered. In early 1990 a limited number of prototype houses and subsequently a large (12 000 m²) factory were built using X-concrete panels imported from Sweden. The factory panels were 8 m high and were used as infill cladding units in a reinforced concrete frame. Although the new system was heralded enthusiastically and received some coverage in the technical press, the trading company did not survive for long. It is not known whether this was due to difficulties of panel supply or whether finally the necessary compliance certificates for the United Kingdom were not forthcoming. It appears at the present time that this is perhaps not an area where new products are likely to achieve a major breakthrough despite their ready acceptance on the Continent.

Again in this section a few case studies are included, generally dating from the late 1960s and early 1970s. In some cases these could almost equally well be added to those in section 5.3.1 and vice versa, as they may be buildings where only a part of the structure is precast.

1. The Student Union Building at San José in California is an interesting example of a lightweight aggregate concrete structural frame designed for an area subject to seismic activity. It comprises a lower ground floor level and three storeys above ground; for the latter all structural elements and walls are of lightweight aggregate concrete, some precast and prestressed. Precast, prestressed T-units are used for the floors; they were cast off-site and their low weight was a benefit for both transportation and erection. They span up to 24 m to provide a large clear space. All internal and external surfaces are of exposed concrete, with a board-marked finish. An expanded shale aggregate, 'Basalite', was used for both the coarse and fine fractions, together with natural sand [9].

**Figure 5.12** Scotstoun House: precast wall unit. (Arup Associates, Architects, Engineers & Quantity Surveyors.)

2. Scotstoun House at South Queensferry, designed by Arup Associates, is one of the buildings investigated in the survey by Mays and Barnes. It is single-storey building constructed to a fast programme in 1966.

   The external walls are loadbearing. They are constructed from precast concrete units which alternate with tall fixed windows, with primary roof beams spanning between the units (see Figure 5.12). The units are U-shaped in plan and section, with a deep reveal to the spandrels and a shelving recess provided internally. Lightweight aggregate concrete was selected for its thermal properties and low weight: the concrete is exposed, and the units were precast off-site. Foam slag aggregate was selected on a somewhat experimental basis and a dye was added to give a tone similar to the local stone used in the surrounding garden walls.

   The survey noted the concrete as being 'generally in good condition', although some minor spalling and horizontal cracking was observed. Spalling caused by low cover has led to the replacement of some cill units.

3. A residential building was constructed in Melbourne, Australia, during the late 1960s using lightweight concrete for all wall panels, both

loadbearing and non-loadbearing, and for the floor slabs for 30 levels above the ground floor. Loadbearing wall panels and the floor slabs were precast and partly prestressed. The use of lightweight aggregate, in this case an expanded shale 'Shalite', facilitated the transportation and erection of large panels [9].

4. Precast cladding units incorporating lightweight aggregate were used for the National Theatre in Tokyo in the mid-1960s. The units are strongly profiled with horizontal bands and a coloured concrete was used to produce a traditional appearance. The aggregate used was 'Lionite', an expanded shale.

5. The Lloyds building at Chatham has roof units in the form of truncated pyramids (a similar form was used for Gateway One at Basingstoke). The units are approximately 7.2 m square on plan and were cast on site. The use of lightweight aggregate concrete gave obvious weight advantages for site handling and erection (see Figures 5.13 and 5.14).

*(a) Standard building systems.* The use of lightweight aggregate concrete for standard precast units appears to be more widespread overseas than in the United Kingdom.

Reference was made earlier in this section to the development of lightweight aggregate concrete wall panels in the former Soviet Union.

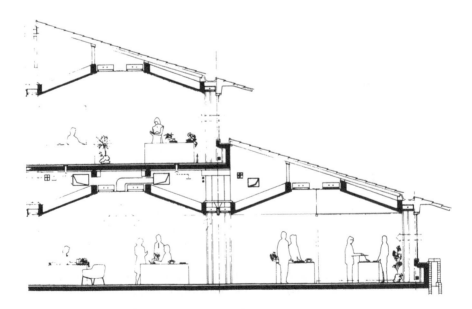

**Figure 5.13** Lloyds, Chatham: typical office section. (Arup Associates, Architects, Engineers & Quantity Surveyors.)

**Figure 5.14** Lloyds, Chatham: internal view. (Arup Associates, Architects, Engineers & Quantity Surveyors.)

Here low-cost systematised construction using a number of standard, mass-produced concrete units for industrial and, in particular, residential buildings is virtually the norm. The range of panels produced is quite extensive; for particular locations (regions and sometimes individual towns) a series of panels will be designed using local materials as far as possible. Figure 5.15 shows the urban planning model of the new residential district for the right bank of the river Dnieper in the city of Kiev. A large number of base systematic units allows the creation of housing complexes varying in configuration and number of storeys, and satisfies demographic requirements that differ within urban territories. The six base systematic units used in a specific design series for this residential district are illustrated, together with the elevation of 'Housing Complex No. 1', in Figure 5.16.

An addendum to the equivalent of a British Standard covers the Design of Residential Buildings [18]. The use of large size structural components

**Figure 5.15** Urban planning model of new residential district in Kiev. (Department of Systematic Designs of the State Design Institute for Civil & Industrial Construction, Kiev, Ukraine.)

is recommended for buildings constructed of prefabricated concrete elements, and the construction of external walls using single-skin, double-skin and triple-skin panels is covered in some detail.

Components are classified as wall panels, roof slabs, floor slabs and three-dimensional blocks. The latter covers small self-contained units and also the equivalent of the prefabricated 'pod' which has come into vogue in recent years but is more associated with steel-framed construction.

Recommendations are given for external and internal wall thicknesses (multiples of 25 and 20 mm respectively), for floor slab spans (up to 4.5 m is 'small span', up to 7.5 m is 'medium span') and for the spacing of stability walls, which should not exceed 24 m. The degree of standardisation becomes apparent from data such as this. Joint tolerance values are also given for the connection of floor slabs to wall panels, dependent on the method of erection of the latter. There are rules for the connection of roof and floor slabs to bracing elements where they are required to provide diaphragm action to resist horizontal forces, and the vertical surfaces of slabs are required to have keyed joints. Given the level of guidance implied, it is not surprising to find that there are numerous standard details covering, *inter alia*, wall panel joints, wall panel/floor slab joints and arrangements for shear force transfer in vertical wall joints.

For external lightweight aggregate concrete walls the following aggregate types are recommended:

*Coarse aggregate*: keramsite, perlite, aggloporite, slag pumice, blast furnace granulated slag and porous natural aggregates.

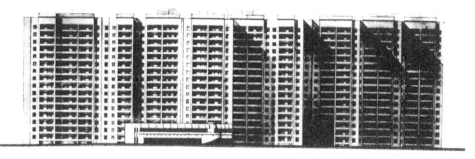

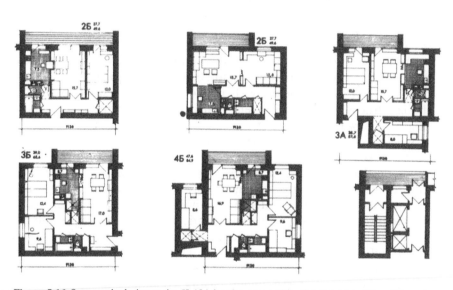

**Figure 5.16** Systematic design series K-134 for the new residential district in Kiev. (Department of Systematic Designs of the State Design Institute for Civil & Industrial Construction, Kiev, Ukraine.)

*Fine aggregate*: crushed keramsite, 'distended' perlite sand, ash and ash with slag mixtures.

Similarly there are recommendations for large reinforced concrete wall panels for industrial buildings. This form of construction is noted as being 30–40% less labour intensive than equivalent brick cladding, is lighter and requires significantly less steel fixings. Quality control is also better.

Recommendations are given for buildings with and without heating systems. Multi-layer and solid panels are used for the former and ribbed panels for the latter; standard dimensions are given for panel size, rib size, spacing, etc. Conventional and prestressed reinforcement is used. Lightweight aggregate concretes with a variety of aggregates are used.

Panels are available in various sizes up to 12 m in length, and special pier panels are made for use with non-systematised or continuous windows. The range allows variety in the arrangement of glazing adopted, using discrete areas or continuous bands. Corner panels are used to complete the wall construction. Panels may be loadbearing or non-loadbearing used with steel or concrete frames. Again, there are standard details for connections and prescribed joint widths depending on the length of panel. Panel to column connection details are carefully considered, needing to cater for differential movements due to thermal effects and frame settlement. The fixings are generally located to allow access for inspection and repair if needed; this is of particular importance where the internal air humidity is high. A variety of lightweight aggregates is used in the manufacture of these panels with a range of density from 900 to 1200 kg/m$^3$.

The extent of standardisation is demonstrated by the existence of state design organisations employing up to 2000 people engaged solely in the design of prefabricated housing. It is also noted that a whole branch of the construction industry was engaged in the implementation of these designs in new housing districts.

Similar systems exist in Germany and, albeit only over the last few years, in Australia.

For the latter, a range of precast wall and flooring panels is manufactured using BST aggregate, small polystyrene beads/granulate which are specially treated and coated during production. This material is also used to produce cladding panels and a variety of elements such as internal partition walls, floor planks and bridge decking. A similar range, to produce a total building system, is manufactured using autoclaved aerated concrete.

In Germany all components for domestic housing and industrial buildings can be manufactured under factory conditions using lightweight aggregate concrete. Liapor manufactures a range of units which includes lintels and even precast window frames for construction from cellar to roof. Floor units are designed generally for imposed loads of 2.7–3 kN/m$^2$. Solid loadbearing wall panels for buildings up to four storeys high are allowed by DIN 4232 [19]; these are storey-height panels which are (relatively) light to handle and offer good thermal performance. The lower modulus of elasticity and heat absorption (around 20% less than for normal weight concrete) reduce creep and shrinkage stresses. Surface crazing, however, tends to occur in precast elements cast face down because of the accumulation of material at the base.


### 5.3.4 Blockwork

The development of the lightweight aggregate concrete blockwork market in different countries is governed in part by the availability and ease of extraction of natural aggregates, and also by pressure to recycle waste materials. In France less than 1% of block production uses lightweight aggregates, despite the advantages, whereas the great majority of blocks in Norway are lightweight, presumably a result in part of the thermal insulation benefits. This section looks at the market in the United Kingdom and then examines briefly the position in some different countries overseas.

Concrete blocks using different lightweight aggregates, and also auto-claved aerated concrete, are now manufactured in a range of densities to produce blocks in varying combinations of strength and both thermal and sound insulation. In addition, such blocks are produced with a variety of surface finishes and thicknesses and are available in some cases with an integral layer of insulation on the face.

Savings in both time and material cost are obtained as handling is easier due to the reduction in weight, the workmanship content is reduced as the blocks are larger (and therefore fewer units are required when compared with brickwork) and there are savings in insulation.

*(a) Strength.* Blocks manufactured with lightweight aggregates are available generally in strengths of 2.8, 3.5 and $7.0 \, \text{N/mm}^2$. Strengths of 4.2 and $6.0 \, \text{N/mm}^2$ are also produced by some manufacturers and 10.5 may be available to order.

Density increases and thermal performance decreases for the higher strengths. The correlation between strength and density is obviously similar to that for concrete mixes described earlier. Dry densities are in the range $480-1450 \, \text{kg/m}^3$, with blocks at the higher end of the range being strictly regarded as medium-dense but still some 30% lighter than traditional dense concrete masonry.

*(b) Use.* Blocks may be used in both internal and external walls; for the latter they are commonly used as the inner leaf of a cavity wall, but they may also be used as the external leaf and tiled/rendered. Use in external walls is limited to some degree by the low characteristic flexural strength [20], for all values of compressive strength, particularly for masonry which spans vertically. The low self-weight of the masonry results in little modification to the flexural strength for cases where the design vertical load is low, e.g. walls where timber or precast concrete floors span parallel to the wall.

Medium-dense blocks with a minimum compressive strength of $3.5 \, \text{N/mm}^2$ are used with precast prestressed concrete T-beams to provide a rapidly laid, efficient and economical suspended floor suitable for domestic use. Proprietary systems have been developed relatively recently, in response to the mandatory requirement of the NHBC [21] that ground floor slabs must be suspended where the depth of infill exceeds 600 mm. These incorporate standard blocks produced by the

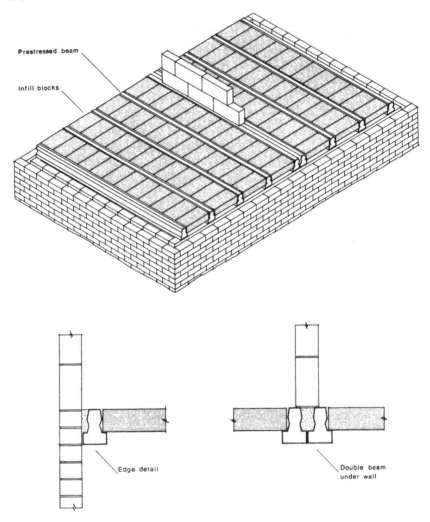

**Figure 5.17** Typical floor system using lightweight concrete blocks.

major manufacturers and provide a fast and economical working platform which avoids the need for costly site preparation. (Typical details are shown in Figure 5.17.) These provide insulation benefits at ground floor, although at upper levels denser blocks are required to meet the requirements of the Building Regulations [22] in relation to sound insulation.

Lightweight blocks are available to a variety of specifications suitable for use below dampproofing course level. Here the use of solid walls incorporating lightweight units offers savings in labour and time when compared with cavity brick construction. The blocks also provide a significant level of perimeter insulation, reducing the degree of ground floor heat losses.

Lightweight hollow and cellular blocks which may be used in reinforced masonry construction are also manufactured.

For comparison, a brief section is included here on the use of concrete masonry in the former Soviet Union, which is a vast country with different climatical and time zones, landscape and environmental conditions. The design of all structural elements, selection of building materials, application and marketing strategy for construction industry products has to take into account the varying climatological and geophysical factors.

Zoning of the ex-Soviet territory based on average temperatures of coldest/warmest months of the year, on the weight of the snow cover, and wind pressure values is given in SNiP 2.01.07–85 [23]. (Note: this document is still in force in ex-Soviet republics.)

Among the variety of structural elements used for civil and industrial construction, masonry occupies a substantial place. Walls, piers, flat slabs, vaults, arches and lintels are designed and constructed from different types of bricks and blocks. Sometimes they are used in combination with reinforcement as reinforced masonry, or with reinforced concrete elements as strengthening components. Dense and lightweight concrete blockwork, with the latter generally falling into the medium-dense category (1200–1500 kg/m$^3$ – see Table 5.1) is widely used.

For all types of masonry the outdoor temperature of the coldest month is an important factor in determining the wall thickness of the buildings. Typical wall thickness values and required thermal resistance values for the range of outdoor temperatures −10°C to −50°C are shown in Table 5.1.

It is of interest to note that lightweight natural stones (weight not exceeding 32–40 kg per stone, rock density <1800 kg/m$^3$) are also used here. The relevant data is included for comparison.

Very large block sizes are used to achieve an improved technical

**Table 5.1** Required thermal resistance (R) and external wall thickness (b) values for residential and public buildings dependent on the outdoor temperature

| Design temperature of the outdoor air (t, C°) | | −50 | −40 | −30 | −20 | −10 |
|---|---|---|---|---|---|---|
| Required thermal resistance (m$^2$ × hr × C°/kcal) | | 1.50 | 1.32 | 1.13 | 0.94 | 0.76 |
| Type of masonry | Density (kg/m$^3$) | Wall thickness (b, cm) | | | | |
| Solid concrete blockwork | 1500 | – | 69 | 59 | 49 | 39 |
| As above | 1200 | 69 | 69 | 49 | 39 | 29 |
| Concrete blocks used for blockwork ('Peasant' trade mark) | 1500 | – | 59 | 49 | 39 | 29 |
| Shell limestone masonry | 1200 | 69 | 59 | 49 | 39 | 39 |
| Tuff used as masonry | 1600 | – | 69 | 59 | 49 | 39 |

**Table 5.2** The three block types, with typical dimensions

| Course | Corner | Lintel |
|---|---|---|
| | | |
| Length (mm): 990 | As course blocks | 5990 mm |
| 1490 | length plus thickness of | |
| 1990 | the wall | |
| 2290 | | |
| Height (mm): 585 | As course blocks | 585 mm |
| 1185 | | |
| 1785 | | |

performance and better economy. Blocks used are course, corner or lintel-type, depending on their location (see Table 5.2).

Block thicknesses are 300, 400 or 500 mm, and lintel-type blocks are reinforced. It is recommended that the external surface of the blocks is covered with 30–50 mm of 'decorative' concrete screed, although it is not clear whether this is solely for aesthetic reasons or also for protection. Blocks are laid on ordinary cement mortar with tuck pointing, but, interestingly, vertical joints are filled with lightweight aggregate concrete, presumably using fine aggregate. Reinforcing bars of 8 and 10 mm are used in joints at the corners of the buildings and where vertical joints coincide. The blocks are tied back to the columns in framed buildings using special T-shaped anchors formed from 10 mm bars [24].

In Australia the BST lightweight aggregate described in section 5.3.3 is used for the manufacture of lightweight blocks. It appears that these have a growing market, particularly in Sydney and Brisbane; for loadbearing applications this is generally for low-rise construction. It is perhaps worth noting here that, as for precast panels, blocks made from autoclaved aerated concrete (AAC) have also been introduced fairly recently, with similar use and application as the BST blocks; standard blocks have a compressive strength of $3.5 \, N/mm^2$ but higher strengths can be produced.

German blockwork practice is circumscribed by regulations, as is apparent from the considerable number of standards (DINs in Germany) which exist. Dense concrete blocks are not widely used but lightweight concrete, incorporating a variety of aggregates, and aerated concrete are common. In the early 1990s the German market for lightweight aggregates was approximately one and a half million cubic metres a year, with two and a half million cubic metres of natural pumice used. Over half of the lightweight aggregates are used in masonry blocks and it is interesting to note that a substantial proportion is used in the manufacture of lightweight mortar; this material is not common in the United Kingdom but its use in Germany is a recognition of the disproportionate heat loss which occurs through the joints in a wall. U-values have been brought down to 0.4 or even less using these materials. Lightweight concrete solid and hollow blocks are regulated by DIN 18151 and 18152 [25]. For fair-faced blockwork a variety of sizes and types of block are available. Both solid and hollow blocks are manufactured and, as in the United Kingdom, there are blocks which are suitable for use below dampproof course level. At least one manufacturer, Liapor, manufactures coloured blocks for fair-faced work.

Compressive strengths generally of 2.5, 5.0 and 7.0 N/mm$^2$ are available. Blocks may be used in single-leaf walls, double-leaf fair-faced walls, rendered walls or cavity construction. (They can also be spray-painted, the latter on a rough or relatively smooth surface as required.)

*(c) Construction aspects.* While there may be overriding reasons for specifying medium or dense concrete blocks, there are obvious advantages in terms of handling in using blocks at the lighter end of the range. Work will be faster with lighter lifting equipment for the lighter materials and manual handling also becomes possible. There are now specific regulations concerning the lifting of heavy loads on site.

At a meeting of the Health and Safety Commission's Construction Industry Advisory Committee (CONIAC) in 1990, the following advice was agreed unanimously: 'Bricklayers should not be expected to lift blocks weighing more than 20 kg. Unless it is known that lifting aids will be available during the laying operation, then architects, consulting engineers and quantity surveyors should avoid specifying such heavy blocks.'

In addition, the European Temporary or Mobile Worksites Directive which covers, *inter alia*, minimum health and safety requirements for the manual handling of loads where there is a risk, particularly of back injury, to workers came into effect in June 1992. The Directive, which is all-embracing but lacking in quantitative detail, puts the responsibility onto the employers of the workmen at risk. However, this responsibility may be widened when the Directive is translated into UK law. A consultation document on the new rules, the Construction Design and Management Regulations (CONDAM) was issued in July 1992 for the industry's response.

### 5.3.5 Refurbishment

The use of lightweight aggregate concrete in refurbishment may take advantage of any of the forms described in the previous sections. Reference was made in the case studies of *in-situ* concrete structures to One Shell Plaza Tower in Houston, where 17 additional storeys were constructed using lightweight aggregate concrete on similar foundations to those for the lower, normal weight concrete option. The advantages in refurbishment again often relate to the construction of replacement/ additional floors in an existing structure while avoiding the need for any strengthening to the foundations.

One good example of this in the United Kingdom was the construction of the London International Financial Futures Exchange ('LIFFE') within the central courtyard of the Royal Exchange Building in London in the early 1980s. The Royal Exchange was built in 1842 and is listed as an Ancient Monument; the structural solution for the new LIFFE construction had to minimise any weight imposed on the old masonry vaults and avoid cutting into the existing fabric. A composite slab using lightweight aggregate concrete on metal decking was selected for the new mezzanine structure to achieve a lightweight solution. The new slab, approxmately $500 \, m^2$ in area, was poured over one weekend. Almost $100 \, m^3$ of concrete was poured in three sessions, requiring 23 loads of concrete and believed to be the longest pumping operation for the material at the time (this volume has now been greatly exceeded, as illustrated in 5.3.1 and 5.3.2) [26]. The Lytag mix was pumped a distance of up to 80 m from the readymix trucks. The slab was power floated.

Another potential use of lightweight aggregate concrete in refurbishment is for improving the seismic resistance of structures with, for example, timber floors. Lightweight aggregate concrete has a greater resilience and lower vibration frequency than its normal weight counterpart. It also has a lower modulus and a higher strain capacity, which increase its resistance to seismic forces. Its low density allows the potential of upgrading a structure without imposing significant additional load on the existing vertical elements and foundations.

Lightweight aggregate concrete infill has been used in bridge refurbishment, including a disused Victorian railway bridge; for the Herne bridge at Oundle, Lytag concrete was used to reduce the dead load over the original brick arches in its conversion to a road bridge for the new Oundle Bypass [27]. Similar technology may be applied to the conversion of old warehouse buildings.

A different example in this instance incorporating Boral Ytong lightweight concrete (autoclaved aerated concrete) was constructed in the early 1990s in Sydney. Three additional floors were added to a four-storey brick and timber building constructed at the beginning of the century. Site and design constraints included retention of the existing façade and re-use of the existing foundations, with restricted site access. This was achieved using structural steelwork and 'Supafloor' slabs between 150 and 200 mm in thickness and 1–6 m in length. Four hour fire-rated

'Supapanels' were used for cladding and lightweight blocks made of the same material were used for all internal walls. Floor slabs were reinforced for the specified loading; generally they are suitable for imposed loads of around $4\,kN/m^2$, but here in local areas a load intensity of up to $8\,kN/m^2$ was required.

## 5.4 Economics of lightweight aggregate concrete in building

### 5.4.1 Introduction

Despite the fact that lightweight aggregate concrete has been used structurally for more than two decades in many parts of the world, and for substantially longer in North America, there remains a reluctance among designers to specify it, particularly for *in-situ* construction; there is also a feeling that it is expensive. The former is based to some extent on inertia, although its technical properties are now well documented and its advantages recognised. The latter arises essentially from a lack of detailed data on both actual construction costs and costs in use and is therefore a response to the rather limited information available. This shows, correctly, that the basic price of the raw aggregates is higher than the gravel equivalent and that the price of readymixed lightweight aggregate concrete supplied to site is also higher than its 'standard' counterpart.

The difference in the basic price has been a consistent factor tending to reduce the use of lightweight aggregate concrete over a long period of time. Also for equivalent strength a lightweight aggregate concrete mix has a higher cement content than the normal weight counterpart; this too is reflected in the cost. Twenty years ago, during construction of the NLA Tower in Croydon, lightweight aggregate were noted as being some 15% more expensive than gravel in London, with a corresponding premium in the price for concrete delivered to site (a Lytag coarse and fines mix).

Ten years ago a study found Lytag granular was 80% more expensive per tonne than 20 mm gravel (although the cost of the aggregates per cubic metre of concrete was almost identical), with a price differential of around 30% for $30\,N/mm^2$ readymixed concrete using natural fines. Typical prices for readymixed concrete in Central London in 1992 show a similar addition for lightweight aggregate concrete, with the percentage increase tending if anything to be higher, particularly for $35\,N/mm^2$ concrete.

### 5.4.2 The Concrete Society study

The Concrete Society published as a technical paper [28] in September 1983 a report by one of its working parties comparing the cost of a multistorey office block designed in both lightweight and normal weight (gravel) structural concrete.

For both materials flat slabs with waffle and solid floor construction were compared for a fairly conventional building in central London. In addition to comparing the cost analysis for the actual construction (capital cost), costs in use were compared as far as possible.

At around the same time the second edition of the *FIP Manual of Lightweight Aggregate Concrete* was published [29]. Both publications arrived at broadly similar conclusions when comparing the economics of lightweight and normal weight structural concrete.

In summary, they found that the construction costs of the two materials were broadly similar, while if costs in use were also taken into account there was a slight advantage for lightweight aggregate concrete. These general conclusions were arrived at by investigating basic material costs, costs of the actual construction process and such costs in use as were both comparable and quantifiable.

The Concrete Society study was intended to extend and underwrite the findings of an earlier (1972) comparative cost study [30] which looked at concrete floor slabs constructed with lightweight aggregate concrete and normal weight concrete (the latter was referred to in the report as 'dense' concrete). Rather than being solely an elemental study, a more comprehensive design and costing exercise was carried out for the later report.

*(a) Technical specification.* The building selected was a high quality speculative office development of nine storeys including a single basement in central London; the gross floor area was $5076\,m^2$ (nett lettable area $4108\,m^2$). Below ground all structural elements (excluding columns) were designed in normal weight concrete. For all other elements, excluding the top-floor penthouse which was of steel-framed construction, normal weight and lightweight concrete alternatives were designed. Piled foundations were assumed.

The design was to CP 110 [31]. The imposed load for the office floors was $(4 + 1)\,kN/m^2$ including partitions. Unit weights for concrete were taken as follows:

| | |
|---|---|
| lightweight concrete (reinforced) | $19\,kN/m^3$ |
| normal weight concrete (reinforced) | $24\,kN/m^3$ |

The lightweight concrete contained Lytag coarse aggregate and natural fines, and 370 kg OPC per $m^3$, an increase of $40\,kg/m^3$ when compared with the normal weight concrete mix. A $30\,N/mm^2$ mix was specified.

A typical floor measured approximately $39 \times 15.2\,m$ with bays $6.0 \times 6.6\,m$ and a projecting edge strip of approximately 1.5 m. A solid flat slab 230 mm deep was used, with a 305 mm deep waffle slab considered as an alternative, using standard moulds and a topping 80 mm deep. The additional thickness of topping avoided the need for shear reinforcement in the ribs. Overall stability was provided by the lift core and stair walls. Cavity construction was specified for the external walls (brick and lightweight block) as part of the general building specification.

*(b) Cost comparison.*   Capital costs were compared for the entire building using the two different concrete mixes and both forms of slab construction, i.e. four sets of costs were obtained. Elements common to each (services, finishes, external works, etc.) were costed on the basis of conventional quantities. Owing to the limited availability of data, manhours and site times were calculated using the same rates for both dense and lightweight concrete; this was believed to be biased towards the former and it was thought that reductions in construction time leading to a saving in capital cost of approximately 3% could be achieved using the latter.

For both solid and waffle slabs, savings in capital cost of 0.2% were identified when using lightweight concrete. Thus the higher cost of the lightweight concrete itself is more than offset by savings in foundations, formwork, fixing of services (lightweight aggregate concrete is easier to cut and drill), reinforcement and handling.

The study also noted various savings on costs in use, which gave further advantages when considering lightweight aggregate concrete but were not in general quantified because of lack of available data. For example, lightweight aggregate concrete improves thermal insulation and reduces heating costs, but no calculations were available. Similarly, the material has a better performance in fire but this benefit was not recognised in terms of reduced insurance premiums, although an estimate was made of the saving for reinstatement.

### 5.4.3   Other information

Detailed comparisons similar to the Concrete Society study – and even that inevitably had to use a number of assumptions – had not been carried out in the United Kingdom for buildings until relatively recently. The conclusions reached are broadly borne out by the findings of a number of studies both in the UK and in America relating to bridge construction, but well-documented factual information from completed jobs does not appear to be readily available to support a possibly rather more 'theoretical' view.

A feasibility study was carried out [32] for one of the Docklands office buildings surveyed in the BCA publication *Economic Long-Span Concrete Floors* [7]. The study was based on preliminary drawings for the scheme and considered two alternative structural solutions to the normal weight concrete solid flat slab construction of the original scheme. It is acknowledged that other options could also have been included but were discounted on the grounds of time limitation.

The building comprises 13 storeys above ground at approximately $1200 \, m^2$ per floor, with two basements. Typical bay size is $8 \times 7.2 \, m$. Piled foundations and a diaphragm wall are utilised. In carrying out the study the possible savings in foundation costs due to reduced load are investigated; thereafter basement construction is taken as the same for each option and frame costs are evaluated for comparison. The options investigated are:

1. post-tensioned construction ($40 \, N/mm^2$ concrete, $225 \, mm$ flat slab for normal weight concrete, $200 \, mm$ for lightweight aggregate alternative);
2. trough slab construction using both normal weight and lightweight aggregate concrete.

The broad conclusions are:

1. the post-tensioned floor slab using metal shearheads and high-strength concrete columns offers cost advantages;
2. using lightweight aggregate concrete floors, there is a further saving based on budget costs.

Savings are identified in three areas:

1. substructure (reduced piling costs)
2. frame
3. cladding (reduced storey height).

Programming, with particular reference to the minimum floor cycle, and details relating to the construction such as column standardisation, type of formwork, pump capacity and line length, are considered in estimating the savings. A four-day floor cycle, possibly reducing to three, was considered feasible. This was to be achieved using unbonded tendons and stressing in a one-stage operation by applying a stress level at 36 hours in excess of the full required value. Despite identifying savings of approximately £600 000 (of which a third was in the foundations) the feasibility study was carried out at a relatively late stage and there was insufficient confidence externally in the technology to proceed with the recommended solution.

A further feasibility study was carried out for a smaller building in Corsham Street [33]. The site was acquired with planning consent but development has been delayed. The brief for this was to obtain a building with high-quality finishes to a budget of £55/ft$^2$ for structure, services and cladding.

The building is $14 \, m$ wide and $96 \, m$ long, with accommodation on four storeys above ground and no basement. Standard institution office loading of $(4 + 1) \, kN/m^2$ has been assumed.

For this building three alternatives were investigated. Guidelines on buildability to reduce costs and reduce programme time while optimising on quality were adhered to in developing these alternatives.

(a) *Lift slab–flat slab construction*

This utilises a 275 mm post-tensioned flat slab, using 40 N/mm² light-weight aggregate concrete. The slab spans 10 m between columns with 2 m cantilevers beyond the column line. Columns are precast using a 60 N/mm² mix and spaced at 6 m centres longitudinally. Stability is provided by frame action longitudinally and by masonry wall panels in the central and wing core areas in the transverse direction. The floors are cast on the ground and lifted hydraulically into position. The core areas are constructed after the slabs are in position.

(b) *Precast composite lift slab construction*

The structure comprises precast columns at 6 m centres supporting an *in-situ* composite floor cast at ground floor and hoisted into position hydraulically once minimum strength has been achieved. The floor slab consists of precast, prestressed beams acting compositely with a slab of precast and *in-situ* construction. The beams at 3 m centres span 11 m over a supporting edge beam, with 1.5 m cantilevers. Concrete grades are 60 N/mm² for the columns, 50 N/mm² for the precast beams and slabs and 25 N/mm² for the edge beams and *in-situ* topping. Stability is provided as for option (a) above.

(c) *Double-T wall and floor units*

This scheme uses precast concrete loadbearing double-T wall units supporting 2.4 m wide precast double-T floor units which span the full width of the building. The double-Ts are standard precast, pre-stressed units constructed with 60 N/mm² concrete. An *in-situ* topping (25 N/mm²) is cast on the floor units. Loadbearing masonry panels in the central and end core areas provide stability in the permanent condition.

For all three alternatives, the servicing of the building and the cladding were considered, and construction sequences for the structure were pre-pared. Finally, cost and programmes were estimated. The programmes were produced with consideration of appropriate levels of labour and plant requirements and costs calculated accordingly.

|            | Cost                        | Programme                     |
|------------|-----------------------------|-------------------------------|
| Option (a) | £706 000  (£127.4/m²)       | 12 weeks                      |
| Option (b) | £718 000  (£129.5/m²)       | 12 weeks                      |
| Option (c) | £1 185 000  (£213.9/m²)     | 9 weeks                       |
|            |                             | (+15–20 weeks lead time)      |

The programme includes an allowance for site set-up.

Costs are inclusive of preliminaries, insurance, contingency and profit.

It can be seen from the above that, subject to a number of assumptions and exclusions necessarily made for a study of this nature, each option has been costed well within the established budget. The costs for options (a) and (b) are significantly less than those for option (c), with the *overall* programme also being shorter (time *on site* is less for option (c)). Of the two, the post-tensioned flat slab using lightweight concrete is again the most economical.

Other studies have been carried out in this country for air-rights buildings. A number of dramatic buildings were erected in central London in the 1980s over main railway stations. While lightweight concrete has been used extensively here, it has almost exclusively been in thin slabs acting compositely with metal decking in steel-framed buildings. A combination of the use of lightweight aggregate concrete and post-tensioning techniques could be considered as a competitive alternative to steelwork for buildings such as these with long-span transfer structures.

### 5.4.4 The selection of lightweight aggregate concrete

As mentioned earlier, a number of economic factors affect the decision whether to use lightweight aggregates. There are both positive and negative aspects, which may perhaps be divided into three categories:

(1) design-related costs
(2) construction-related costs
(3) costs in use.

It is significant that nearly a decade on from the Concrete Society study, which found both capital cost savings and advantages in costs in use when using lightweight concrete, the vast majority of *in-situ* structures in the United Kingdom are still constructed using normal weight concrete. At the same time there is a certain irony in the significant increase in the use of lightweight concrete which has occurred during the same period; this of course is associated with the proliferation of composite construction and steel-framed buildings, where lightweight concrete has been widely specified for some of the very advantages that were identified in the study.

### References

1. Holm, T. A., Bremner, T. W. and Newman, J. B., Lightweight aggregate concrete subject to severe weathering, *Concrete International*, Detroit American Concrete Institute, June 1984.
2. BS 8110: Structural use of concrete: Part 2, Code of practice for special circumstances, British Standards Institution, London, 1985.
3. Lightweight Aggregate Concrete: Designers' Fact Sheets (to BS 8110). Prepared by the Concrete Society's Lightweight Concrete Committee, *Concrete* (May/June 1991).

4. Mays, G. C. and Barnes, R. A., The performance of lightweight aggregate concrete structures in service, *The Structural Engineer*, **69** (No. 20; 15 October 1991), 351–61.
5. Spitzner, Dr Ing Joachim, High strength lightweight concrete, *ERMCO Congress, Stavanger, Norway*, June 1989.
6. Lytag in sliding formwork, *Lytag Technical Bulletin*, No. 10, Boral Lytag, November 1970.
7. Matthew, P. W. and Bennett, D. H. F., *Economic Long-Span Concrete Floors*, British Cement Association, 1990.
8. *The Sky's the Limit. A Century of Chicago Skyscrapers*, Rizzoli International Publications Inc, New York, 1990.
9. CEMCON, *Lightweight Aggregate Concrete. Technology & World Applications*, Associazione Italiana Tecnico Economica del Cemento, Rome, 1974.
10. Mould, G., Tower block development, Guy's Hospital London, *The Structural Engineer*, Volume **49** (No. 1; January 1971), 35–53.
11. CP 114: 1957: The structural use of reinforced concrete in buildings, British Standards Institution, London.
12. *Modul 2 1970*, Modul Verlagsgesellschaft GmbH, pp. 4–23.
13. Zunz, G. J., Some examples of when and how lightweight concrete should be used, *International Congress on Lightweight Concrete, London*, May 1968.
14. *Construction Weekly* (19 August 1992), 17.
15. Precasting in the Soviet Union, *Precast Concrete* (May 1981), 225.
16. *Construction News* (No. 6152; 1 March 1990), 20–21.
17. *Building Today*, **199** (No. 5839; 18 January 1990), 18–20.
18. *Manual for the Design of Residential Buildings*, Issue 3, Structural Elements of Residential Buildings (Addendum to SNiP 2.08.01–85).
19. DIN 4232: Technical Building Regulations 6/88, Deutsches Institut für Normung.
20. BS 5628: Part I: 1978, Structural use of unreinforced masonry (reprinted 1985), British Standards Institution, London.
21. *NHBC Standards Chapters 5.1 and 5.2*, National House-Building Council, 1990.
22. *The Building Regulations 1991*, Department of the Environment and The Welsh Office. Approved Document E: Resistance to the passage of sound, HMSO, 1992.
23. SNiP 2.01.07–85: *Loading (Loads & Exposure)*, State Committee for Construction of the USSR, Moscow, 1987.
24. Dyatkov, S. V., *Industrial Buildings and their Structural Elements*, Visshaya Shkola, Moscow, 1971.
25. DIN 18151: Lightweight concrete hollow blocks, 9.87, Deutsches Institut für Normung. DIN 18152: Lightweight concrete solid bricks and blocks, 4.87, Deutsches Institut für Normung.
26. *New Civil Engineer* (No. 511; 7 October 1982), 28–29.
27. 'Bridge building', *Concrete* (June 1988), 20–22.
28. *A Case Study of the Comparative Costs of a Building Constructed using Lightweight Aggregate and Dense Aggregate Concrete*, Concrete Society Technical Paper 106, The Concrete Society, 1983.
29. *FIP Manual of Lightweight Aggregate Concrete*, 2nd edition, Fédération Internationale de la Précontrainte, 1983.
30. *A Comparative Study of the Economics of Lightweight Structural Concrete Floor Slabs in Building*, Lightweight Concrete Committee Cost/Price Working Party, The Concrete Society 1972 (revised and updated, 1978).
31. CP 110: Part I: 1972, Code of practice for the structural use of concrete, British Standards Institution, London.
32. Bennett, D., *Arrowhead Development Preliminary Report*, March 1988.
33. Bennett, D., *Corsham Street Development: A Preliminary Study of the Concrete Frame Options for a Four-Storeyed Office Building*, British Cement Association, September 1990.

# 6 Lightweight concrete in bridges

J. H. J. MANHOUDT

## 6.1 Introduction

Lightweight concrete has been used in a number of heavily loaded concrete bridges for a long time now. A Concrete Society report in 1981 reviewed the use of lightweight concrete for bridges in a number of countries [1]. The report concluded that the higher initial cost of the concrete had been more than offset in savings in reinforcement and prestressing steel, reduced construction costs and simpler piers and foundations. Potential savings of 10% were identified as being possible. The report also highlighted the potential for the use of lightweight aggregate concrete in the upgrading of existing bridges.

The use of lightweight aggregate concrete for bridges has been fairly limited in the UK. The Concrete Society report lists only six. More recently the access bridge to the car park of the Ealing Broadway Centre in London, a cable-stayed structure with a clear span of about 33 m, was built in lightweight aggregate concrete having an average strength of over 40 N/mm$^2$.

## 6.2 Why use lightweight concrete in bridges?

Lightweight concrete as delivered to site is more expensive than normal concrete. The material is more expensive because the lightweight aggregate is more expensive than gravel, broken stone, etc. The ever-increasing energy costs are a big handicap for sintered products. Moreover, lightweight concrete requires a higher cement content.

The use of lightweight concrete requires more care. Because lightweight concrete transports energy less easily, more attention has to be paid to its compaction. Vibrating has to be done in a closer pattern, but for less time at each point. Pumping sometimes presents problems that require extra precautions, especially when a lot of concrete is cast. (This is discussed in more detail in chapter 4).

However, these disadvantages are offset by the advantage of the lower weight. Depending on the type, a concrete bridge's own weight is around 70% of the total load. Using Lytag, for example, around 20% of the

Table 6.1 A comparative study of materials for the Koningspleijbrug

|  | Normal concrete (gravel) | Lightweight concrete (Lytag) |
|---|---|---|
| Concrete | 28 800 m³ | 28 565 m³ |
| Reinforcement | 2 900 tonnes | 2 760 tonnes |
| Prestressing | 730 tonnes | 680 tonnes |
| Piles | 19 700 m | 17 000 m |

bridge's own weight is eliminated, or around 15% of the total load. Using Liapor aggregate the figure may be as much as 20% of the total load. This all results in a lower construction depth, less reinforcement and prestressing and a lighter foundation.

However, these advantages are only effective if the lower weight of the material is taken into account throughout the design. A reduction in the construction depth is only of significance if the bridge can be built lower so that it has lower piers and shorter access ramps.

A comparative study of the Koningspleijbrug (a bridge), at Arnhem in the Netherlands, which will be discussed later, indicated a considerable saving in the use of materials (see Table 6.1).

Tenders were invited for construction in both a normal concrete and a lightweight concrete, and the lightweight concrete version won the race, although the difference was small, in the region of 1%. However, it must be remembered that this bridge was the first to be constructed in lightweight concrete for a long time and the aggregate used was new to the Netherlands.

The Lightweight Concrete Committee of the Concrete Society also carried out a study of the efficiency of lightweight concrete bridges [2]. This study was conducted on a bridge with a 25 m span. The design was based on the use of prefabricated 1200 mm deep beams with a cast *in-situ* decking (see Figure 6.1).

The dimensions of the beams in the design using normal dense concrete were the same as those for the lightweight version. The version using

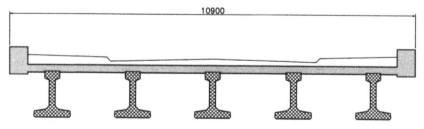

**Figure 6.1** Cross-section of lightweight concrete bridge deck used in design study [2].

normal concrete required six beams, whereas the version using lightweight concrete needed only five. The cost saving for the decking was 1.44%. The saving for the whole bridge was 0.3%. This small, but nevertheless positive, saving was achieved even with this relatively small span.

## 6.3   Types of aggregates used for bridges

Various lightweight aggregates were used in bridge building in the past. The advantage of a material with a low density appeals to the imagination.

Korlin and Liapor are older types of aggregate. They are made of clay, formed into balls and then sintered. These sintered types use a lot of energy in the production process, and the ever-increasing energy costs in the past made these aggregates less and less attractive to use. Liapor has been used for several major bridges; for example, the Sandhornoya Bridge in Norway, and, in the Netherlands, the Rottepolderplein Bridge near Haarlem, the bridge crossing the river IJssel near Deventer, the bridge crossing the river Meuse near Well and the Dukenbrug Bridges near Nijmegen.

A type of lightweight aggregate that is presently gaining favour uses another basic material, namely fly-ash (or pulverised-fuel ash, PFA). Examples include Lytag and Aerdelite. Lytag is a sintered product while, in the case of Aerdelite, a binding agent is used. The advantage of Aerdelite is that the production process requires less energy. However, a disadvantage is that the particle is not as strong as the Lytag particle. This means that a stronger type of concrete can be made using Lytag. With a Lytag mix a concrete strength of $60 \, \text{N/mm}^2$ can be reached easily. For Aerdelite the maximum is about $30 \, \text{N/mm}^2$. Lytag is thus the more interesting for use in bridge building.

It still requires a lot of energy to produce the Lytag particle although the remaining carbon in the PFA provides the majority of the energy required. However, the basic material is a waste product. The PFA is obtained from flue gases in coal-burning electricity generators. The PFA contains a number of harmful substances, such as heavy metals. Consequently, PFA has to be treated as a contaminated material. When disposing of PFA, measures have to be taken to prevent it entering the environment. The sintered PFA grain has a ceramic skin that safely envelopes the harmful substances.

Environmental considerations mean that, in some countries, PFA has a negative value, i.e. the aggregate manufacturer is paid to remove the ash from the power station. Thus Lytag may nevertheless be competitive as far as price is concerned, despite the high energy costs.

Lytag particles are heavier than Korlin or Liapor particles and the gain, as far as density is concerned, is therefore smaller. However, the properties of Lytag are closer to those of normal concrete and effects such as fragmentation, brittle fractures, etc., are less likely to occur.

## 6.4 Advantages and disadvantages

The following examination of the advantages and disadvantages of light-weight concrete is based on the use of Lytag, as this is the most important type of concrete for bridge building in the Netherlands.

The most obvious advantage is its low density. Whereas normal concrete has a density of $2400 \, kg/m^3$, the figure for Lytag concrete is $1950 \, kg/m^3$ for the same compressive strength. This results in the use of less material in the support structures and foundations. The amount of material saved depends on the type of construction for which the concrete is used. A few examples are given in section 6.7.

At the Koningspleijbrug, a bridge at Arnhem, the large span produced the saving. The bridge across the Sinigo, at Bolzano, also benefited from the short side spans. These side spans consist of 14 m long ballast blocks. The lightweight central span reduces the weight of the ballast blocks.

Another advantage of the light weight becomes apparent when hoisting and moving prefabricated bridge girders, deck, plates, etc. The work can be done using a crane with 20% less lifting capacity, or the structure can be divided into fewer, larger precast units.

The round particles make Lytag easy to work with. There is less wear on mixers and formwork. This, again, is discussed in more detail in chapter 4.

Lytag particles and cement mortar have almost the same density, which means that there is hardly any separation during casting.

The thermal-insulation capacity is three times higher than that of normal concrete. Moreover, Lytag particles have a higher resistance to splitting at high temperatures. Therefore, a building element remains intact longer in a fire.

Lytag concrete is almost unaffected by hydration cracks. If a wall made of normal concrete is cast on a previously cast floor, cracks occur at about every 5 metres. These cracks arise on account of the heat of hydration which causes a considerable temperature increase in the concrete when it is still plastic. The warm concrete subsequently hardens and then cools. This causes what are known as shrinkage cracks when the concrete is restrained; the term 'heat of hydration cracks' would be more appropriate. If Lytag concrete is used, this phenomenon hardly occurs at all because the material has

- a low thermal expansion coefficient;
- a low modulus of elasticity;
- a large elongation at rupture.

In the case of a box girder, where the bottom plate has already been poured, the pour sizes used for the walls can be bigger; and when pouring safety barriers on a deck, the pour sizes can again be bigger.

Cooling of the concrete is therefore generally unnecessary when casting thick sections using lightweight concrete. If cooling is employed, the high thermal resistance must be taken into account. The sphere of influence of cooling pipes is therefore relatively small.

Lytag concrete is also readily drilled and nailed. Alterations and extensions can therefore be made easily; for instance, it is easy to drill holes for railing anchors, etc.

A difficult to quantify, but nevertheless important advantage, is the environmental effect. Lightweight grains made of PFA result in the efficient and long-lasting use of residues and contribute to solving the problem of where to store these substances. Furthermore, it is an answer to the environmentalists' increasing resistance to the use of natural resources such as gravel, rock, etc.

Besides these advantages there are also disadvantages. Lytag particles are not as strong as natural aggregates such as gravel. If a fracture occurs, the plane of the fracture continues through the particle, as can be clearly seen in Figure 6.2. Because the particle fails to stop the formation of the crack, the crack will extend. Cracks in lightweight concrete are almost always large and, for this reason, fragments of concrete may also break off. In short, lightweight concrete is more fragile and less robust.

Finishing can be more difficult and requires a special skill. When making circular movements it must be ensured that the grains are not drawn out of the fresh mortar, as this would create a less dense surface. That would be detrimental to the resistance against frost, de-icing salts

**Figure 6.2** Splitting of concrete blocks, using Lytag (left) and normal weight aggregate (right).

and damage by chemicals, although the resistance of lightweight concrete is better than that of normal dense concrete.

## 6.5 Recent research

A lot of research has been carried out into lightweight concrete in the past. The first studies were on Korlin and Liapor concrete, and these were followed by studies on Lytag and Aerdelite concrete.

The different methods of preparing concrete in different countries present difficulties in carrying out research. For example, in the United Kingdom, a cement content of $500 \, kg/m^3$ is not unusual. An example of this is the bridge at Redesdale, which is described in section 6.7. In the Netherlands, for example, a cement content of $350 \, kg/m^3$ is normal. In the Koningspleijbrug (a bridge at Arnhem), mixtures of $360 \, kg/m^3$ were used and strengths of $40 \, N/mm^2$ were achieved. This means that test results cannot be readily compared and used as the cement content may affect them.

Between 1988 and 1992, extensive research was carried out in the Netherlands into the properties of Lytag and Aerdelite concrete. In addition to determining the strength and mechanical properties, such as the cylinder strength, thermal expansion and deformation, studies were also carried out to measure the forces acting on the concrete. This refers to such items as measuring the tensile and compressive stresses, shear stresses, the bond behaviour of reinforcement bars and durability.

All the studies showed that the behaviour of Lytag concrete is very close to that of normal concrete. For instance, the cracking behaviour – crack lengths, crack widths and tension stiffening – is no different from that of normal concrete [3]. The splitting behaviour is also comparable [4]. The behaviour is, however, different with regard to the shear strength, which is decidedly lower [5]. The same is true for overlapping bars where the bond strength is considerably lower. Futhermore, relatively wide cracks will appear if the bars are not staggered at the ends. Cracks caused by splitting were noted just before reaching the maximum load. Shortly thereafter, pieces of the concrete cover broke off exposing the bars along the length of the overlap. The use of stirrups prevents the cover from coming loose. However, there was no question of there being any loss of rotational capacity. A greater lap length or the staggering of the bars solves this problem [6].

The studies will lead to a Dutch standard for lightweight concrete. This standard will be published in 1993 as a recommendation of the Dutch committee of research (CUR), a division of the Dutch Concrete Society.

The studies have also shown that every type of lightweight concrete has its own requirements. Making a link between the properties and the density is sometimes an oversimplification of the relationship.

The density is generally considered to distinguish lightweight concrete from normal dense concrete, and concrete below $2000\,kg/m^3$ is therefore considered to be lightweight, but this is not completely correct. It is relatively simple to make Lytag concrete with a density in excess of $2000\,kg/m^3$. After doing so, it would be possible to treat it as normal concrete. This would ignore the specific properties of lightweight concrete, such as its lower robustness, the fact that the fracture continues through the grain, its lower shear resistance and its different strength and mechanical values. It would be much more appropriate to refer to it according to its constituents, i.e. Liapor concrete, Lytag concrete and so forth.

## 6.6  Recommendations for applications in bridges

When using lightweight concrete in bridges, its different characteristics must be taken into account. These have been laid down in various regulations. Nevertheless, lightweight concrete behaves differently from normal concrete. It may readily be used in bridges, provided that this different behaviour is taken into account. However, its different behaviour can only be partially laid down in rules on, for instance, its reduced shearing-force capacity, the need for longer lap lengths, etc. Because lightweight grains are weaker than natural heavy-aggregate grains, cracks when they occur tend to go through the grain. Since this means that the grain fails to stop the crack, it will continue. Measures have to be taken in those parts of the construction where hair cracks can be expected to occur, for example, because of accidental tensile forces. Small amounts of extra reinforcement, placing the prestressing units a little further from each other and so forth, provide simple and effective solutions.

The measures to be taken in relation to normal concrete may be summarised as follows.

Where there is likelihood of splitting tensions occurring, the upper and lower layer of reinforcement must be joined. For example, this would be the case with construction joints. In a construction joint there will never be a uniformly distributed stress level owing to the compaction of the concrete and shrinkage. Adding some steel to connect the reinforcement layers at the top and the bottom of the concrete element is a good solution to this problem – a diameter of 8 mm at a spacing of 300 mm often being sufficient. These tensions may also occur in the concrete walls between prestressing ducts. When the ducts are blown clean, considerable tensile strains can occur in the concrete between the ducts. Prestressing cables must be 10% further from one another than they would be for normal weight concrete, and must be anchored with larger anchor plates.

It must be assumed that the maximum shearing stresses to which the concrete may be subjected are around 15% less than for normal weight concrete. This is discussed in more detail in chapter 3.

Laps in reinforcement must be staggered and be approximately 20% longer than in normal weight concrete.

However, the most important factor is, and will continue to be, that the designer must engross himself in his construction and ensure that reinforcement is fitted wherever tensile stresses can occur, even accidentally. A small amount of reinforcement can work wonders.

## 6.7 Examples of bridge structures

### 6.7.1 Koningspleijbrug, a bridge near Arnhem, the Netherlands

*Introduction.* East of Arnhem, a new bridge has recently been built over the river Rhine [7]. This bridge will carry the traffic between the south and east of the Netherlands around the city. This will decrease the intensity of the traffic over the bridge in the centre of Arnhem, the famous Arnhem Bridge, well known in the Second World War. Traffic prognoses predicted that 34 000 vehicles a day would make use of the new bridge after completion.

Construction works started at the beginning of 1985. The first part was finished by the end of 1986. The whole bridge was opened for traffic by mid-1988.

*Main dimensions.* The main span is 133.430 m since navigation does not allow for piers in the river. The approach spans over the flood plains are, beginning at the abutments, 37.000, 4 × 49.000 and 80.535 m on each side. Details are given in Figure 6.3.

The width of the bridge deck is 28.185 m, composed of four traffic lanes and a separate bicycle and pedestrian lane.

*Cross-sections.* The main and two adjacent spans are a box section with variable depth. The approach spans over the flood plains have a double-T cross-section. Details are given in Figures 6.4(a) and 6.4(b) respectively.

During design of the approach spans a comparison was made between a double-T and a box section, both with a constant depth.

The box section was planned to be built following the 'pushing' method, the double-T to be cast *in-situ*. The results of the studies between the two possibilities showed minor differences; both solutions were tendered for and the double-T was chosen.

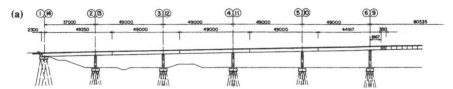

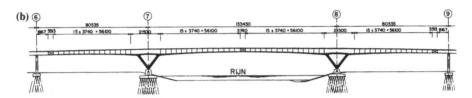

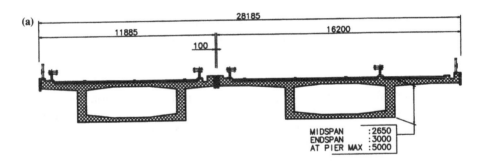

**Figure 6.3** Koningspleijbrug spans: (a) over flood plains; (b) over river.

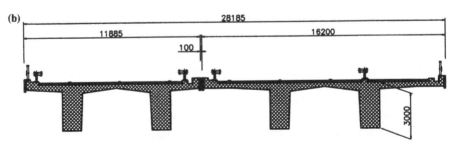

**Figure 6.4** Koningspleijbrug cross-sections: (a) over river; (b) over flood plains.

*Lightweight concrete.* Also during design, a comparative study was made between normal concrete using gravel as aggregate and lightweight concrete with sintered PFA, Lytag, as aggregate. Financial, as well as environmental reasons led to the decision to choose lightweight concrete for the whole bridge, superstructure as well as piers and abutments.

The above-mentioned study comprised the determination of the characteristics of the material, Lytag. Part of the data had been obtained from Pozzolanic Lytag. In the United Kingdom, this product has been applied for many years, though not for major bridge structures with dimensions as here.

It was decided to continue the study of Lytag. Since the usual mix designs in the Netherlands are markedly different from those in the United Kingdom, further evaluations were carried out at a Dutch Government approved laboratory as described earlier. These investigations were related to mix design, compressive and tensile strengths, Young's modulus, creep and shrinkage. They were the basis for the final decision to accept Lytag as the construction material for the bridge. With a mix ratio of 62% Lytag, 38% sand and 360 kg Portland cement, a 28-day strength between 40 and 50 N/mm$^2$ was obtained. The advantages in comparison with ordinary concrete are:

- Economy in weight                    20%
- Linear coefficient of expansion      83%
- Young's modulus                      80%.

The reduced modulus results in greater prestress losses and more deflections. However, if this is taken into account in the design process, it will give no problems.

### Construction method.

*Midspans.* These were built as free cantilevers, cast *in-situ*, starting from the main piers. These piers are V-shaped and consist of two inclined legs connected at the top by a 21.500 m long part of the box girder. On the flood plain side, this part cantilevers about 3.5 m. The operation started from this platform. The segments were alternately added in such a way that, at the flood plain side, the segments preceded. Figure 6.5 shows construction under way. Measurements were taken to make sure that the unbalanced moment was always taken by a support on the flood plain side. No temporary support on the other side was therefore necessary.

*Approach spans.* The approach spans were cast *in-situ*, span by span, with casting joints at one-quarter of each span. The shuttering was a tunnel mould that, after release, could be moved longitudinally and re-

**Figure 6.5** Construction of Koningspleijbrug.

**Figure 6.6** Koningspleijbrug approach spans.

used several times. Figure 6.6 shows one series of spans. The longitudinal prestressing tendons were coupled at the casting joint.

The crossing's double T-sections are deep and generated high temperatures during setting. Cooling pipes were built in, but this was only a

**Figure 6.7** Completed Koningspleijbrug (photo: Aerofoto Brouwer-Brummen).

precaution. The critical temperature rise in lightweight concrete can be much higher than in normal concrete because the thermal insulation capacity is higher. However, such hydration temperatures are acceptable. As explained in section 6.4, cracks will hardly ever occur.

*Details.* The whole bridge is prestressed in the longitudinal as well as the transverse direction.

The bearings of the main piers are rubber bearing pads, placed underneath the inclined legs; all the others are sliding bearings.

The only expansion joints are at the abutments, so the whole bridge is continuous over its total length. Using normal dense concrete, the expansion joints at the abutments would require a large movement capacity. Figure 6.7 shows the completed bridge.

Owing to current concern over environmental pollution, an increasing application of Lytag aggregate can be expected as PFA is a waste product of a power plant and is harmful to the environment. The Lytag factory (VASIM) is situated beside the Nijmegen power plant at about 15 km distance from the bridge site.

*Acknowledgements.*

Bridge client: Gelderland Province, Department of Roads, Bridges and Ground Affairs.
Consultant: BVN Consulting Engineers, Rijswijk.

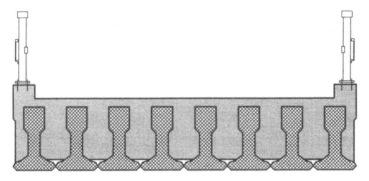

**Figure 6.8** Cross-section through Redesdale Bridge.

Contractor: Pleycom, a joint venture between Dubbers-bouw, Malden en
  Gebr. Haverkort, Vroomshooop.
Contract price: DFL 35 000 000 (excluding VAT).

### 6.7.2  Bridge at Redesdale, UK

The lightweight aggregate concrete bridge at Redesdale is a bridge carrying
a forest road over a river. The bridge has a single span of 16.760 m and a
carriageway width of 3.56 m. Details of the cross-section are given in
Figure 6.8. The deck consists of precast prestressed concrete beams with
*in-situ* infilling and surfacing concrete. Lytag lightweight aggregate was
used for both concretes. The abutments are constructed in a normal
dense concrete.

The mix of the precast concrete contained Lytag coarse aggregate and
local river sand. The cement content was 575 kg/m$^3$ and the calculated
wet density was 1890 kg/m$^3$. The compressive strength at 28 days was
49 N/mm$^2$.

The mix for the *in-situ* concrete contained lightweight aggregates only.
The cement content was 500 kg/m$^3$, the calculated wet density 1760 kg/m$^3$
and the compressive strength 38 N/mm$^2$. For nearly 12 years the strains in
the precast prestressed beams were measured. Load tests on the bridge
were made during the early summer of 1970 and again in 1981. A full
report of the tests is given in the TRRL Supplementary Report 788 [8].
Among other points, the report notes that there is no indication of any
lack of durability including the lightweight concrete deck, which did not
have the usual protection waterproof layer.

### 6.7.3  Bridge in Ringway (ring road) near Ulft, the Netherlands

As part of the re-routeing of the S57 provincial road around Ulft, which
was completed in 1991, two bridges were built using lightweight concrete.

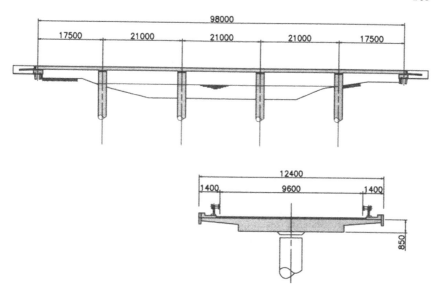

**Figure 6.9** Details of bridge over Oude IJssel, near Delft. Top: longitudinal section; bottom: cross-section.

The first bridge, which crosses the road, is for pedestrians and cyclists. The second is a bridge that carries the road over the Oude IJssel river. The pedestrian and cyclist bridge is a skewed slab with spans of 9.30–14.70 and 8.95 m. The angle between the axis of the bridge and road is 63 degrees. The width is 4.35 m, the depth 0.50. The bridge is of reinforced Lytag concrete.

The bridge over the Oude IJssel is built entirely of Lytag concrete; this includes both the road surface and the abutments. The piers are made of steel pipes. The reason for using Lytag was two-fold:

(a) economy
(b) environmental considerations.

The bridge is 98 m long, 12.400 m wide and is divided over five span lengths. The cross-section of the bridge is a slab 7.100 m wide with overhangs, each of which extends 2.500 m. Details are shown in Figure 6.9. The piers are the special feature of this bridge. They consist of a column at the centre of the bridge. The bridge bearings at the abutments provide the torsional stability.

During the design period, special attention was paid to the bridge deck at the location of the piers. The bearings create large shear forces in the deck and shear forces are not the strongest feature of lightweight concrete. The cross-section at the location of the piers was therefore reinforced. The reinforcement takes most of the shear stresses.

Lengthwise, the bridge is prestressed by 30 cables 12 × 12.9 mm. The piers are made of steel pipes with an external diameter of 1500 mm and a wall thickness of 25 mm. The pipes were driven down to the firm bearing layer. Once in position, the piles were filled with normal weight concrete, which was the only normal weight concrete used in the bridge.

### 6.7.4  Bridge over the river Sinigo at Avelengo (Bolzano), Italy

The bridge is located in the route between Merano and Avelengo and crosses the canyon of the river Sinigo at about 1250 m above sea level.

Since the total abutment-to-abutment span to be crossed was 125 m, the most economic solution would have been a continuous beam on four supports (two abutments and two intermediate piers, the subspans being 31, 63, 31 m), but this solution ran head-on into the difficulty of accessing the canyon's sides, both because of the steepness of the slopes and because no service road could be built without having to cut down numerous good-sized trees along its route. The inevitable choice was therefore a single 125 m long span.

The structural solution was a triple hinge arch, shown in Figure 6.10. Furthermore, during construction the two semi-arches had to act as cantilevers until the hinge at the top could be set in. It was thus necessary to provide a temporary counterbalancing and anchoring system at the abutments, able to take the high loads that a 62 m long cantilever would transmit. It was obvious that just the dead weight of the abutment blocks would not suffice to balance the weight of the cantilevered semi-arch, or at least not with an adequate safety coefficient. On the one hand, the structure could be anchored to the rock by temporary prestressed tie-

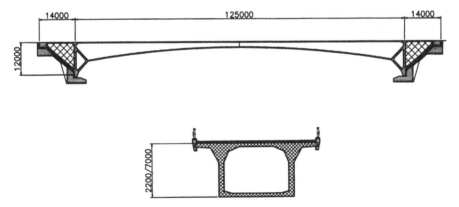

**Figure 6.10** Details of bridge over river Sinigo at Avelengo. Top: longitudinal section; bottom: cross-section.

bars; and on the other, the dead weight of the advancing structure could be reduced. The second approach was adopted. The segments were cast using expanded-clay concrete, with a lower density than normal concrete. The abutments functioned during the construction phase as counterweights.

The abutments are truss shaped. They were cast in normal dense concrete and filled with massive lean concrete to increase the balancing moment.

Since a large number of cables were anchored in the end of the counterweight, there was a high local stress concentration. This led to the requirement for a series of transverse prestressing rods. The total bridge length is 158 m, of which 125 m is clear centre span. The total deck width is 8.10 m.

The cantilever structure is formed from a single box girder 5 m wide at the bottom. The depth of the box girder is 2.20 m at mid-span and 7 m near the abutments. Each cantilever consists of 15 precast elements. Further details of the bridge are given in reference [9].

### 6.7.5  The Friarton Bridge in Scotland

The Friarton Bridge in Scotland was completed in 1988 and has nine spans varying from 63 to 174 m. This steel box girder bridge was originally designed to have a mastic asphalt surface laid directly on an orthotropic steel deck.

Design changes to meet the Merrison design rules led to the adoption of a 100 mm thick reinforced concrete deck acting compositely with the steel through shear studs welded to the top flanges of the boxes and secondary steelwork.

By using lightweight concrete it was possible to avoid redesigning the piers and foundations. Because of the very critical density requirements, Lytag was used as coarse and fine aggregates. The aggregates were transported by rail over a distance of about 700 km to a local readymixed concrete plant close to the bridge site.

Despite repeated attempts with different mixes and sources of supply of aggregate, it was not possible to achieve nominal air dry cube densities down to the design value of 1680 kg/m$^3$ at 28 days for a concrete having a mean cube strength of about 38 N/mm$^2$. The mix finally chosen had nominal densities of 1690–1750 kg/m$^3$. Final densities of the concrete as placed on the bridge varied with weather conditions and the time of year at which the deck waterproofing was laid. This effectively seals in any moisture in the concrete and prevents further moisture movements. Some difficulties were experienced initially with maintaining adequate workability due to absorption of water after batching. This was overcome by adding extra water on site for up to 1½ hours after batching. Tests showed that this could be done with no apparent detrimental effect on strength.

## 6.7.6  Refurbishment and upgrading

There have been a number of cases in which lightweight aggregate concrete has been used in the refurbishment and upgrading of bridges. The reduced self-weight has enabled an increased live load to be carried by the supporting structure. In some cases the deck has been widened, allowing extra traffic lanes.

One example is Blackfriars Bridge in London [10]. Increased loading standards required either a weight restriction or a substantial upgrading of the bridge. The first option was unacceptable, leading to the requirement to strengthen the substructure or reduce the deadweight. The deck had to be replaced anyway, at a cost of about £1.2 million. The additional cost of using lightweight concrete was about £84 000, compared to £890 000 for strengthening the substructure, making lightweight the obvious choice. Holm [11] gives brief details of a bridge over the Hudson River in the USA, which was widened from two lanes to three lanes. Using a lightweight concrete that weighed about 25% less than normal concrete meant that less than 20% of the steel support structure needed strengthening. Similarly, the Woodrow Wilson Memorial Bridge in Washington, DC [12] was widened using precast lightweight concrete units, giving an additional width of 1.8 m to each carriageway.

## 6.8  Summary and conclusions

This chapter has shown that lightweight aggregate concrete can be used economically for the construction of bridges. However, as discussed in the case studies, it is essential that the reduced density, and the changed properties, be taken into account at all stages, both in the design process and when considering the construction.

## References

1. *A Review of the International Use of Lightweight Concrete in High Bridges*, Concrete Society Technical Report No. 20, August 1981.
2. *Design and Cost Studies of Lightweight Concrete Highway Bridges*, Concrete Society, Special Publication, 1986.
3. Delft University of Technology, Faculty of Civil Engineering. Stevin Laboratory and TNO Institute for Building Materials and Structures. Report 25.5-91-3/VFC, June 1991.
4. Delft University of Technology, Faculty of Civil Engineering. Stevin Laboratory and TNO Institute for Building Materials and Structures. Report 25.5-91-2/VFC, June 1991 and Report 25.5-91-4/VFC, July 1992.
5. Delft University of Technology, Faculty of Civil Engineering. Stevin Laboratory and TNO Institute for Building Materials and Structures. Report 25.5-91-6, May 1992.
6. Delft University of Technology, Faculty of Civil Engineering. Stevin Laboratory and TNO Institute for Building Materials and Structures. Report 25.5-92-5/VFC, July 1992.

7. Manhoudt, J. H. J., *Koningspleijbridge*, FIP Notes 1987/1, pp. 5 and 6.
8. Higgers, G. E., Loe, J. and Howells, H., *The Lightweight Aggregate Concrete Bridge at Redesdale*, TRRL Supplementary Report 788.
9. Segre, E., A lightweight concrete bridge over the Rino Sinigo at Avelengo (Bolzano), *L'Industria Italiana del Cemento* (12/1983), 769–72.
10. Anon, Blackfriars set to clear the deck, *Construction News* (12 May 1988).
11. Holm, T. A., Structural lightweight concrete for bridge redecking, *Concrete Construction* (August 1985), 667–72.
12. Lutz, J. G. and Scalia, D. J., Deck widening and replacement of the Woodrow Wilson Memorial Bridge, *PCI Journal* (May/June 1984), 74–93.

# 7 Lightweight concrete for special structures
## B. K. BARDHAN-ROY

### 7.1 Introduction

Lightweight aggregate concrete has been successfully used in a wide range of constructions – from conventional dwellings and office blocks to complex highly specialised structures.

In general, their performance in service in terms of safety serviceability and durability has been satisfactory – which underlines the suitability of the material for structural application. In fact many of the physical properties/characteristics of lightweight aggregate concrete, such as high strength to density ratio, low thermal strain and conductivity, low permeability, high resistance to impact and thermal shock, high fire resistance, high tensile capacity, etc., make it an obvious choice as a medium of construction. It is particularly suitable for high-technology special structures where, apart from the response to normal gravity loads, containment of natural and environmental forces with directional conflict as well as thermal effects also play a significant role in the design.

The intention in this chapter is to describe and illustrate actual or potential applications of lightweight aggregate concrete in such special and sophisticated structures. Examples include structures completed or under construction, as well as some interesting schemes where the use of lightweight aggregate concrete was envisaged. These schemes were, unfortunately, abandoned or shelved owing to the prevailing climate of economic uncertainties, after the design and development process had been quite advanced.

There is, of course, no clear-cut definitions as such, of a 'special' structure; and, for simplicity, the structures which are ostensibly novel and imaginative in concept and design or, at the frontier of the construction/material technology of the time, are considered here as belonging to that category.

For efficient design, improved serviceability and overall economy, such structures are usually prestressed and use high-strength concrete. With the advancement in the concrete technology high-strength concrete in the range of $40-80\,\text{N/mm}^2$ can now be produced relatively inexpensively with many types of lightweight aggregates.

The application of prestressing to lightweight aggregate concrete is, however, to a certain extent, inhibited by the fact that the design based on the current codes of practice (where guidance is rather inadequate in regard to lightweight aggregate concrete) cannot show worthwhile improvement over dense concrete prestressed design in terms of cost-effectiveness [1]. As a result, only in special circumstances, when some overriding conditions may prevail or when the design procedure can be adjusted to reflect the actual characteristics of the concrete to be used, is the prestressing of lightweight aggregate concrete considered.

In that sense any prestressed lightweight concrete structure can be regarded as a 'special design'.

Some worked out examples of prestressed lightweight aggregate concrete structures have been included in the Appendices to show the design procedures for a bridge deck, for a long cantilever beam and also for high fire resistance on the basis of reference [2].

## 7.2  Concrete quality

It has been stated earlier that to maximise the efficiency and performance of the 'special structures', high-strength concrete is generally needed. Unfortunately, not all lightweight aggregates are suitable for this purpose. On the basis of applications and functions, the FIP Manual [3] classifies lightweight aggregate concrete into three categories: Category I (Structural), Category II (Combined Structural and Insulation) and Category III (Insulation).

Category I concrete, which is for structural application, covers a broad range of density and strength, density (air-dry) usually varying between 1600 and 2000 kg/m$^3$ and the concrete between the strength limits of 20 and 80 N/mm$^2$, or more. The band is so wide that FIP Manual has subdivided this category into three further grades:

Grade I:   Strength range 20–40 N/mm$^2$ with density (air-dry) between 1600 and 1750 kg/m$^3$.

Grade II:   (Medium strength) Strength range 40–50 N/mm$^2$ with density between 1700 and 1900 kg/m$^3$.

Grade III: High strength concrete from 50 to 70 N/mm$^2$ or more with a density up to 2000 kg/m$^3$.

Grades II and III are normally used in prestressed construction and among the lightweight aggregates available in the UK, Lytag (sintered pulverised-fuel ash), Granulex (expanded slate imported from France), Pellite (pelletised blastfurnace slag) and 'Liapor' (expanded shale – imported from Germany) appear to be capable of producing such high-strength concrete in factories or even on sites, with a careful mix design and high degree of quality control. In such mix designs, lightweight fines are usually replaced by sand. Under laboratory conditions, strength up to $100\,N/mm^2$ has been achieved using lightweight aggregates and sand. Using silica fume (as part replacement of cement) and superplasticiser (for reducing the water/cement ratio) strength in excess of $100\,N/mm^2$ can be produced within an air-dry density of $1850\,kg/m^3$. When silica fume is added, lightweight fines need not be replaced by sand.

For pretensioning, a high strength at an early stage is required to maintain a quick production cycle. As in gravel concrete, steam curing can be effectively applied to lightweight aggregate concrete for this purpose. Research actually shows that owing to better insulation properties, lightweight aggregate concrete benefits more from steam curing at an early stage than gravel concrete [4]. Higher temperature than would be permissible for normal weight concrete can be used with consequent reduction in curing time. Ultra-fine cement (specific surface $7500-8000\,cm^2/g$) has also been used to achieve high strength at an early period [4].

## 7.3   Examples of application

Examples of special structures with lightweight aggregate concrete are broadly grouped under two headings: (a) marine and offshore structures and (b) onshore structures.

### 7.3.1   Marine and offshore structures

It was indeed in a marine structure that the modern lightweight concrete (i.e. using manufactured lightweight aggregate) was first applied. In order to save the steel needed for military purposes, pressure was building up towards the end of the First World War to use concrete in building ships to replace the depleted merchant fleets. The feasibility study by marine engineers at that time indicated that a concrete ship would be practical if the concrete used could have strength over $35\,N/mm^2$ and a density less than $1760\,kg/m^3$ [5]. Normal weight aggregates could not satisfy the density requirement, while the available natural lightweight aggregates could not achieve the required strength level. With the collaboration of American Ship Building Authorities, Stephen J. Hayde of Kansas City at that time developed an artificial lightweight aggregate – rotary-kiln-produced expanded shale – which was strong, inert and durable and was able to produce concrete of strength over $35\,N/mm^2$ within the density limit required by the marine engineers and thus considered fit for use in shipbuilding. The manufacture of the aggregate, on a commercial scale, soon followed under the proprietary name of 'Haydite', ushering in the era of the modern lightweight concrete.

During the world war periods (the First World War as well as the

Second) many ocean-going ships were built with reinforced concrete, and despite higher densities (higher than the $1760 \, kg/m^3$ limit suggested by the marine engineers) normal weight concrete was also used in the construction of many of these vessels without much problem. In terms of buildability, strength, safety, durability and even operational qualities, the concrete ships (both normal and lightweight) are found to have a distinct edge over their steel counterparts. Low maintenance, relatively long service life and economic repairability of hull damage make additional plus points for concrete ships [6, 7]. But an unfavourable deadweight/displacement ratio compared to steel vessels often offsets these advantages and virtually precludes the use of concrete in large ocean-going transport vessels [6]. The problem of the deadweight to displacement ratio obviously diminishes in the case of lightweight aggregate concrete, because of its reduced density.

It is, perhaps, of interest to note in this connection that the weight differential between lightweight concrete and normal weight concrete becomes much more pronounced in a submerged state. For example, if in the atmosphere the weight of lightweight concrete is 75% of that of normal weight concrete, in the immersed state this will be reduced to the order of 55%.

It is however, the advance in concrete and construction technology, together with the improvement in prestressing techniques in the post-war period (Second World War), that has actually changed the scenario quite appreciably and made the construction of large transport vessels in concrete (normal weight or lightweight) viable, both from the technical and economic points of view. The combination of prestressing and high-strength concrete can substantially reduce the cross-sectional area of the structural parts in the hull, and by taking advantage of concrete's excellent mouldability can further help achieve a much more favourable and structurally efficient configuration of the hull with complex double curvature or similar form than is possible with steel [7]. The direct consequence of all this is a significant improvement in deadweight/displacement ratio for concrete ships in general and lightweight concrete ones in particular, narrowing the gap in this respect with steel vessels. Using prestressed lightweight concrete the deadweight/displacement ratio can be increased to 70% compared to 80% in the case of a steel vessel [8]. For 100 000 and 200 000 tonne tankers, the ratio 80% is achievable.

The low permeability of prestressed high-strength (lightweight) concrete enhances the durability of the structures made with the material – which is very important for marine or offshore structures where leaktightness is essential.

For all types of marine structures – from offshore storage platforms, floating production facilities for oil or gas or minerals, floating terminals and dry docks to large ships for transportation of cargoes – prestressed concrete is now a competitive and viable medium of construction [6, 8]. Whether the concrete is of normal weight or lightweight aggregate depends mainly on the circumstances and/or the economies of the individual project. However, in special cases such as vessels for transportation of liquified natural gas (LNG) at cryogenic temperature, the application of lightweight aggregate concrete is most effective. The thermal properties, as well as other containment properties of lightweight aggregate concrete such as permeability, strain capacity, resistance to thermal cracking, etc., have been found to be markedly superior (at low temperatures) to those of normal weight concrete [9, 10]. For example, the permeability coefficients are about two orders of magnitude lower, crack resistance is approximately double and, if cracks appear, the widths tend to be smaller [10].

Tests have shown that most of the concrete properties (both normal and lightweight) remain unchanged at cryogenic temperature, and some of them are even enhanced.

Some examples of actual or proposed application of lightweight concrete in such structures are as follows:

*(a) The ship* Atlantus *[5].*   Haydite lightweight concrete was used in the construction of the 3000 ton ship *Atlantus* which was successfully launched in December 1918. As the record goes, it appears to be the first known structural application of lightweight concrete made with manufactured aggregate.

*(b) The USS* Selma *[5].*   After the success of *Atlantus* the 132 m long, 7500 ton reinforced concrete tanker USS *Selma* was built in the following year, also using Haydite concrete (i.e. expanded shale lightweight concrete) (Figure 7.1). It has a place of pride in the history of lightweight aggregate

**Figure 7.1** Launching of USS *Selma*.

concrete and has demonstrated, perhaps more than any other structure, the excellent durability in an aggressive environment. After just over three years of service the tanker was purposely sunk (in 1922) in Galvaston Bay, Texas, where it remained in a partially submerged state. The hull was mainly under the sea water with a band of about 1 m height being exposed to cyclic wetting and drying process, while the upper section was constantly subjected to salt air with occasional sea water splash due to wind or wave action.

After about 31 years the boat was partially recovered and the condition of the concrete studied. Tests on cores taken from the waterline area revealed a very well compacted concrete with compressive strengths ranging between 55 and 75 N/mm$^2$. The modulus of elasticity was found to be about 20 kN/mm$^2$ and a bond strength on plain reinforcement, 3.56 N/mm$^2$. Remarkably no evidence of any reinforcement corrosion could be seen although the cover to steel was only 15–16 mm. In 1980 a further study was made of the state of the structure by testing and examining more concrete cores taken from both submerged and partially exposed areas. No sign of corrosion could be observed nor any evidence of microcracks even when viewed under a magnifying glass. The compressive strength was in the region of 70 N/mm$^2$ and the $E$-value about 25 kN/mm$^2$ [5]. There was no evidence of any deterioration or degradation of concrete after nearly 60 years of exposure to the hostile environment.

*(c) Other ships* [8].    During the period 1918–22, 15 vessels were built in concrete using expanded shale lightweight aggregate concrete. One of these *Peralta* was afloat until recently.

The concrete shipbuilding programme virtually came to a standstill after 1922 and did not resurface until 1942/43 (in the middle of the Second World War) when shortage of steel again forced governments of many countries to reintroduce concrete in ship construction. In the USA, 104 ships were built using expanded shale lightweight aggregate – during the second world war period – including SS *David O Saylor*, the 5200 ton tanker which was launched in 1943.

In Germany, and also in the UK, experiments continued with concrete vessels and other floating structures. The 2425 ton *Lady Kathleen* (Figure 7.2) was one of the two coasters built in Britain using expanded clay lightweight aggregate [8]. The barge, built in 1943 was in service until 1951 when it hit a sandbank in the Baltic and was grounded.

*(d) LNG (Liquified Natural Gas) tanker.*    As stated earlier, prestressed lightweight aggregate concrete is undoubtedly the most suitable material for construction of LNG transportation vessels. Although none has yet been built, it is evident from available documents that there have been several serious proposals for constructing LNG carrier ships in prestressed

**Figure 7.2** Coaster *Lady Kathleen*.

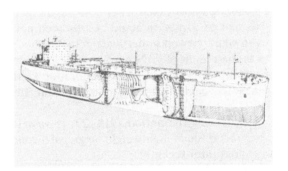

**Figure 7.3** LNG tanker proposed by Dytam Tanker GmbH.

lightweight concrete, some of which were even processed through to an advanced stage of design, after a detailed comparison of technical merits as well as of capital cost, maintenance and operational costs with similar steel tankers.

The most well-known proposal for lightweight concrete LNG tankers (Figure 7.3) is perhaps the one which Dytam Tanker GmbH (a joint venture company of Tampimax Tankers Limited (UK) and Dykerhoff & Widmann (Germany)) intended to build and proceeded through to working drawing stage [11]. The proposal is quite novel in concept.

The spherically curved form of the hull and the bulk head shown in Figure 7.4 is intended to help resist the forces by membrane action, enhance structural efficiency and provide better protection against impact. Unlike the conventional steel LNG carrier, the proposed vessel has the container (i.e. cargo tanker) and the hull constructed as an integral unit,

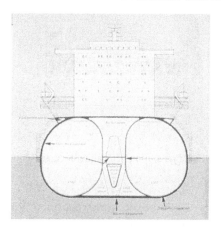

**Figure 7.4** Hull configuration.

i.e. 'single hull' concept [11]. The actual cargo is kept separated from the hull wall by a layer of insulation and some form of primary and secondary barriers in order to minimise thermal shock. The hull is fully prestressed transversely and reinforced longitudinally. High-strength lightweight concrete (pumpable type) of strength $40\,N/mm^2$ and density below $2000\,kg/m^3$ has been specified. Both the prestressing steel and rebars are cryogenic temperature resistant. Other technical data of the vessel are as follows:

| | |
|---|---|
| LNG cargo capacity | $125\,000\,m^3$ |
| Cargo weight | $56\,250\,t$ |
| Weight of the structure | $49\,800\,t$ |
| Shaft horsepower | $40\,000\,hp$ |
| Speed (loaded) | $19.5\,knots$ |

A comprehensive cost comparison between this scheme and a similar (also fully designed) steel LNG carrier confirmed that, apart from the saving of US $15 000 in the construction cost, the concrete vessels would have an overall saving of 18% (in repair maintenance, operational cost, insurance, wages bill, etc.) in 20 years of service. (The comparison was based on 1976–78 prices [11].)

*(e) Floating dock in Genoa Harbour* [12, 13]. This is a very ambitious project and when completed will have a lifting capacity of 100 000 tonnes. At the time of design and development it was thought that it would be capable of containing the largest ships afloat. Some larger ships have, however, been made since. The construction is still proceeding 15 years on; some setback at the initial stage has pushed back the programme for completion substantially.

This Caisson-type dock with an overall length and width of 350 m and

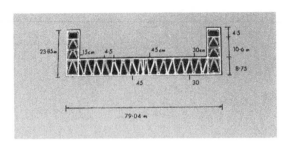

**Figure 7.5** Cross-section of floating dock at Genoa.

78.25 m respectively is a composite construction of tubular space truss encased by prestressed lightweight concrete using expanded clay aggregates. The same type of construction extends to the wing wall which is integral with the dock floor (Figure 7.5).

The total depth of the dock floor is 8.5 m and the height of the wing wall 15.10 m above the dock floor. The length of the Caisson and wing wall is 342.82 m. The width of wall is 6.67 m. The lifting capacity is 100 000 tonnes with an additional reserve capacity of 15 000 tonnes.

The dock is constructed in eight separate equal elements, each 43 m long and 78.25 m wide (i.e. full width). The upper and lower slabs are prestressed both longitudinally and transversely. The successive units are temporarily connected until all eight sections are together, when the whole assembly is joined together by prestressing with post-tensioned cables to make a single integral unit. In addition to analytical assessment using computer, aspects of structural behaviour were also studied by tests on a scaled down model.

It is claimed that the construction cost savings are of the order of 20% compared to a conventional all-steel construction and the reduction in the maintenance cost is about 70%, while retaining the same level of safety and serviceability.

*(f) Underwater oil storage tank.*   The scheme was designed and developed for a contractor who intended to make an alternative bid for the construction of a cluster of seven identical tanks for oil storage, submerged in 150 m of water in the North Sea off the west coast of Scotland. The full project also included a steel tower above the cluster of tanks. The scheme is interesting in that it has an unusual, yet structurally efficient, shape and it is based on a fast erection programme using precast and/or prestressed lightweight aggregate concrete.

The actual time available for the construction of the tanks (cluster of seven) was 11 months only, and to be able to adhere to this time scale the following decisions were made:

1. 'Precasting' should be used extensively.
2. The precast elements should be as large as practical, consistent with the crane capacity and transport facility so that the actual number of units to be erected become as few as possible.

The limitation imposed by the contractor was that the maximum weight of a unit should not exceed 115 tonnes and the maximum dimensions to be within 30 m.

Considering the construction sequence which has been worked out in detail and the stage-by-stage design, six basic types of precast elements were finally decided upon, as discussed later.

Each tank is of a cartwheel shape in plan (see Figure 7.6); the base is a hexagon 45 m across flats and 52 m across apices. The tank has a maximum overall plan dimension of 40 m and it consists of a circular ring 'hub' at the centre with an internal diameter of 10 m. There are 12 radial walls (spokes) 1.25 m thick at the junction with the 'hub' tapering to 0.6 m at the outer end and 12 curved segment walls of 0.5 m thick vaults of 6 m inside radius, connected to the radial walls. The roof over the tanks is 1.2 m thick and extends to the outer faces of the peripheral walls between tanks. The roof will act as a base to the steel tower. The total height of the tank (to top of roof) is 36.4 m.

The seven tanks are connected at the base, at the roof and by infill walls connecting the radial walls on lines between any two adjacent tank centres. The minimum clearance between walls of adjacent tanks is over 5 m.

The overall plan size of the base of the completed cluster of seven tanks is $(3 \times 45) = 135$ m across flats of hexagons $\times$ 130 m across apices. The tanks are set back just over 2.5 m from the flats and just over 6 m from the apices. The overall height is 36.4 m.

The net cubic (storage) capacity of each tank is approximately ($940\,m^2 \times 33.9$ m clear height) $31\,800\,m^3$ and that of the cluster of the seven tanks is $223\,000\,m^3$.

The hexagonal base of each tank, together with the outer walls (curved in plan) up to 6.7 m height above the base, is in *in-situ* concrete using normal weight aggregate. The base is 1.3 m thick. Above that everything is in precast lightweight aggregate concrete.

The number and types of precast elements for each tank are as follows:

1. Element 1 (×25): The central ring or hub unit is 1 m thick, diameter 12 m (outer) and 10 m (inner). Nominal height of each unit 1.340 m.
2. Element 2 (×60 (5 × 12)): The radial wall or spoke unit – each 11.3 m long and 6.7 m (nominal) height with mild steel shoes at top and bottom for vertical connection between successive units. Thickness is

**(a)**

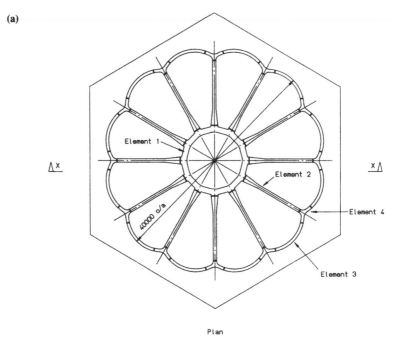

Plan

**(b)**

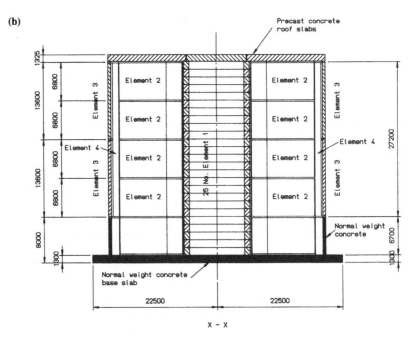

X - X

**Figure 7.6** (a) Plan and (b) section of underwater oil storage tank.

600 mm for 8.5 m length from the outer end and then gradually increases to 1.25 m at the junction with the 'hub'.

3. Element 3 (×24): The curved segment walls or vault units are each 13.6 m (nominal) high, 500 mm thick and 7 m arc length approx. (at outside face) with 6 m radius to internal face. Three mild steel shoes are provided at top as well as at bottom of each unit for vertical connection.
4. Element 4 (×12): The shaped column unit, are each 27.2 m high with three mild steel shoes cast in at the bottom. The units are lightly pretensioned for handling and transport.
5. Element 5 (×1): Circular roof unit 0.6 m thick and 12 m diameter. (Total thickness of slab is 1.2 m so additional 600 mm will be pumped on the precast.)
6. Element 6 (×12): Trapezoidal roof unit 0.6 m thick. (Additional 600 mm will be *in-situ*.)

All lightweight aggregate concrete is in Grade 45.

The choice of lightweight aggregate concrete is dictated by the following considerations:

(a) To limit the weight of the element within the imposed restriction by the contractor without sacrificing the structural efficiency. As a matter of fact the specification requires Grade 45 concrete at an air-dry density not exceeding 1900 kg/m$^3$. The weight of element 4 is approximately 113 tonnes.
(b) Limited float weight depth.
(c) Reduced permeability and improved thermal and insulation characteristics.

It is envisaged that the tanks would be post-tensioned vertically (the hub and the column) and horizontally at shell/column joints, at various stages (see Figure 7.7).

### 7.3.2 Onshore structures

The last three decades have seen application of lightweight aggregate concrete in a broad range of onshore structures, special or conventional – from bridges to buildings and car parks, from grandstands/stadia or sports facility construction to industrial structures, hangers, etc. – in various parts of the world.

In some instances the entire construction is in lightweight concrete while the vast majority of applications seem to be limited to the 'special structural features' only. In every case, however, without exception the choice of the material was made after a thorough comparison of the relative merits and demerits – not excluding the economic considerations – of different alternatives, in the context of the actual project concerned.

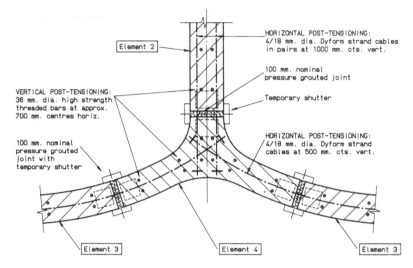

**Figure 7.7** Connection of various precast elements of the tank by post-tensioning.

The particular properties of lightweight concrete which can be important and of direct benefit to onshore structures are low density, high strength/density ratio, high fire resistance and good insulation/thermal characteristics. When these favourable aspects are properly exploited in the design and/or the advantages adequately quantified, the economic justification for use of lightweight concrete in structures becomes relatively simple.

The examples included here cover a wide spectrum of structures (or structural elements), all interesting in design concept and with good record of efficient performance in service.

*(a) Ski-jumping platform at Oberstdorf, Germany.* Built in 1972, this ski-jumping platform is a unique construction combining elegance with technical excellence and blending harmoniously with the environment.

Sited on a mountain top, the structure cantilevers 100 m (measured along the slope) into space at about 39° to the horizontal plane (Figures 7.8 and 7.9) and secured to the base by means of rock-anchors [14]. The entire superstructure is a cellular box section in lightweight aggregate concrete, post-tensioned with Dywidag prestressing bars. The substructure is in normal weight concrete. A lift operates inside to move the skiers to the jumping-off stage platforms up at the free end of the cantilever. Structurally the box cross-section is a very suitable form to resist the severe loading, particularly due to the transverse winds to which the structure is subjected. The construction was carried out using an 'incremental launching' method – a technique commonly used in bridge construction.

The lightweight concrete was made with Liapor expanded clay aggregate

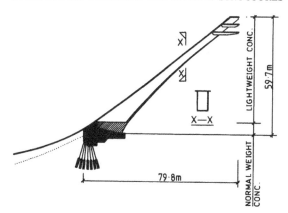

**Figure 7.8** Ski-jumping platform. Oberstdorf, schematic view.

**Figure 7.9** General view of the Oberstdorf ski platform.

having a characteristic 28-day strength of $45\,N/mm^2$ at a density of $1750\,kg/m^3$ [15]. The use of lightweight aggregate concrete resulted in substantial savings in rock-anchors, as well as in prestressing tendons and, consequently, in the overall cost. The structure also shows good weathering properties and good durability of lightweight concrete in a severe exposure condition.

*(b) Bridges.* In many European countries, as well as in the USA and Australia, lightweight aggregate concrete has been used for sometime in the construction of bridges and a number of medium- to long-span major bridges have been built successfully in those countries. The Parrot Ferry Bridge in California, the bridge over the Rhine near Cologne, the bridge over the lake Fuhlingen, Ottmarsheim Bridge over the Alsace canal in France – to cite only a few – are quite well known. Some spectacular foot bridges like the one at Wiesbaden-Schierstein and Bickenstag Villingen are also made in lightweight concrete. It has been proved that when the span is large and the dead weight becomes a predominant consideration in the design, prestressed lightweight concrete becomes very competitive, not only with the normal weight concrete counterparts but also with steel construction.

Most of the bridges mentioned above are multi-span and constructed with the cantilever launching method. Usually the maximum span or the substantial part of it uses lightweight aggregate concrete while the smaller side spans are made with dense concrete for better counter-balancing of the cantilever construction and also to reduce the magnitude of bending moments in the larger span by increasing the stiffness and dead weight in the smaller side spans. The recently completed (1989) bridge on the lower Rhine near Arnheim (Holland) – the Koningspleijbrug – did not, however, follow this system of using different concrete (normal weight and lightweight) in different bays.

The unique feature of this bridge is that the entire construction, including the foundation and substructure (except piling) is made with lightweight concrete using Lytag aggregate. The use of lightweight aggregate concrete resulted in the saving of about 200 tonnes of steel and 2000 m of pile length per phase and in terms of money over £50000.00 per phase [16]. The bridge was completed in two phases of similar construction. The full details of the bridge are given in chapter 6.

In Norway, as in Great Britain, the lack of design guidance and rules for lightweight concrete in the Code of Practice on bridge design did not encourage its application. However, since the introduction in 1987 of the revised Norwegian code of practice, which included design rules for lightweight concrete, there has been a big upsurge in its use for bridge construction, mainly in the superstructure.

The Boknasundet Bridge built in 1990–1991 near Stavanger in the

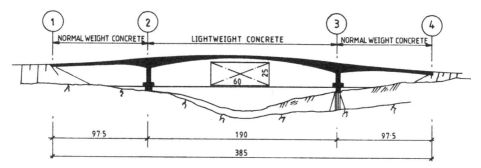

**Figure 7.10** Boknasundet Bridge, schematic view.

South Western region of Norway has probably the largest span (190 m) of all the lightweight concrete bridges in Europe. It has three spans of 97.5 m, 190 m and 97.5 m as schematically shown in Figure 7.10. The central span is in lightweight concrete with Liapor aggregate while the two end spans are in normal density concrete [17]. The characteristic strengths of lightweight and normal weight concrete are 60 N/mm$^2$ and 55 N/mm$^2$ respectively.

The original proposal designed by the Road Department (Owner) was based on normal weight concrete and it consisted of a central span of 150 m with two approach spans of 35 m and 85 m on either side of the central bay. The alternative design with lightweight aggregate concrete which was finally accepted for construction resulted in 6.5% saving from the estimated cost of the owner's original proposal.

Appendix 1 gives design calculations for a lightweight concrete bridge in accordance with the British design code.

*(b) Buildings.*    Many high-rise buildings have been constructed in lightweight aggregate concrete. Two are shown in Figures 7.11 and 7.12 and, as can be seen, each is 'special' in its own right [13, 14]. The third illustration (Figure 7.13) is a shopping centre complex, where a high structural fire resistance requirement for a long-span floor has been satisfied by the use of lightweight aggregate concrete.

*One Shell Plaza, Houston, Texas, USA.*    This 52-storey high-rise office building (Figure 7.11) was built in the late 1960s and, until recently, was the tallest concrete frame building in the world – 215 m high.

The entire superstructure of the building, columns, walls, beams, floors were constructed in reinforced lightweight concrete using expanded shale aggregate. The concrete strength (cylinder) was 30 N/mm$^2$ and the density was 1850 kg/m$^3$.

**Figure 7.11** One Shell Plaza Building, Houston, Texas.

*Central Administrative Building of BMW, Munich, Germany.* Built in 1970 the BMW building is one of the tallest in Munich and has an unconventional shape (Figure 7.12). The structure consists of a central core from which the floors are suspended. The core is in normal weight concrete and the floors – 17 altogether – are built in lightweight concrete using Liapor aggregate. All floors were cast at ground level and lifted to position as the construction progressed. The concrete strength was 25 N/mm² with a density of 1800 kg/m³. By using lightweight concrete in the floor construction, a reduction of load of the order of 13% (average) was achieved on loadbearing members, resulting in a substantial saving on the project cost, without taking into account other inherent benefits of improved fire resistance and thermal insulation.

*Clifton Down Shopping Centre Complex, Bristol.* Built in 1979, the complex comprises a shopping centre, offices and a multilevel car park for the shoppers. The building is divided into three structural units – Blocks A, B and C – each separated from the adjacent block by a nominal 20 mm movement joint (Figure 7.13).

Block A has a basement and three storeys above the ground level and is a conventional reinforced concrete structure with columns at 6 m grids in both direction.

**Figure 7.12** BMW Central Administration Building, Munich.

Block B has a single storey above the ground level, and a basement.

Block C comprises a basement and four split-level parking decks above the ground. The construction consists of precast frames and prestressed double-T floor units with *in-situ* structural topping to provide 15.6 m clear span for customer car parking. The ground floor of the entire complex is the shopping level.

The basement of Blocks B and C is used as a delivery area for the shops (and partly as a car park) and requires 4 hours' fire resistance. For that reason the double-Ts over the basement (in Blocks B and C, spanning 15.6 m and supporting heavy superload) had to be in lightweight aggregate concrete, as it was not possible to achieve such a high fire resistance with normal weight concrete units. In fact, 700 mm deep units in Lytag concrete, with minimum 50 mm structural topping (also in lightweight concrete) was used. Even with lightweight aggregate concrete 4 hours' fire rating could only be proved by analytical assessment in accordance with the procedures in reference [2] – where all detailing conditions and variations could be taken into account. Some additional transverse reinforcement in the ribs of double-T's and continuity rebars in *in-situ* topping over the support region was necessary to achieve this. It is to be noted that if 'tabulated data' in BS 8110 [18] are followed, the double-T construction used may not have fire resistance in excess of 2 hours.

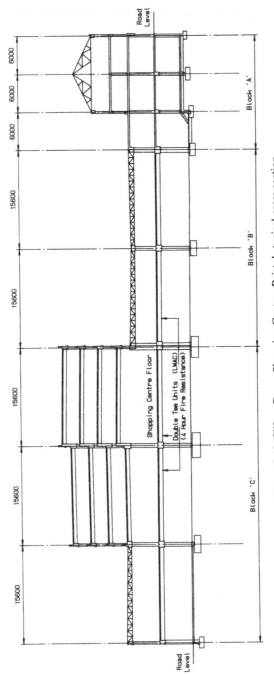

**Figure 7.13** Clifton Down Shopping Centre, Bristol, typical cross-section.

This example of the application of such an analytical assessment is given in Appendix 3.

*(c) Grandstands.* The modern grandstands or stadia aim to provide the spectators with a clear and unobstructed view of the entire arena in which the performance or event takes place. This is essentially achieved by:

(a) making the slope of tier or terrace such that the line of sight of a spectator in one row to the focal point or object is not obstructed by the presence of spectators in the row in front
(b) restricting the distance of the furthest point of the arena (or the playing surface) from any spectator within the 'viewing limit' so that the principle object of the game or event (such as the football, tennis ball, etc.) remains recognisable
(c) keeping the terrace or tier completely column free to avoid full or partial restriction of views.

Usually a roof cover is recommended over the viewing area to protect the spectators from the elements and thus enhance their comfort level to some degree. However, as no column support is available in the area, the provision of any roof often poses challenging problems to the designer, and the larger the extent of the terrace, the more acute becomes the problem, requiring more skill to achieve the solution. This is perhaps why the roof construction is usually the centre-piece or special feature of the grandstand/stadium design.

In the multi-tier stands the difficulty may also arise in the design/construction of the upper tier. Since, for the clear view of lower tier spectators, no column can be taken up to provide the support, these tiers need to be cantilevered. In most cases the extent of the cantilever is quite long and requires special design considerations.

Three examples of grandstands are included here which show how the application of lightweight concrete has led to the solution of the afore-mentioned problems and provided efficient and aesthetically pleasing structures.

*South Stand at Rugby Football Union Ground, Twickenham, UK.* The stand (see Figure 7.14) is constructed on five levels, two of which house hospitality boxes with catering facilities and are independent of the remainder, having separate access stairs and lifts. It incorporates three tiers of viewing areas. The bottom tier, which is separated from the rest of the stand with a longitudinal movement joint, was originally designed for standing accommodation, but has been converted to seating area following Justice Taylor's report on the safety of stadia. The upper two tiers seat approximately 6000 spectators.

**Figure 7.14** South stand for Rugby Union Football, Twickenham.

The interesting feature of the stand is the roof construction. Prestressed lightweight concrete roof beams which are anchor shaped (see Figure 7.15) are suspended from pylons spaced at 13.2 m centres by stainless steel (austenitic) cables. The beams are spaced at 6.6 m centres. Those which are directly on the pylons (i.e. at 13.2 m centres) are supported by two rows of cables in the plane of the beam, while the intermediate ones are picked up by diagonal cables from the pylon.

The roof beams passing through pylons are anchored at the back to the stairflight walls while the intermediate ones are anchored to a vierendeel girder spanning between adjacent stair/lift walls. The beams are prestressed virtually uniformly with 38 No. 12.7 mm diameter Dyform strands. The front 23 m length of the beams are in lightweight concrete with Lytag aggregate, the rear 12 m in normal weight concrete and in solid section to provide increased counterweight. Concrete strength is 60 N/mm$^2$ at 28 days and 40 N/mm$^2$ at transfer. Under the worst combination of service load, the beams are designed as class 3 structures (using the BS 8110 definition), with the hypothetical tensile stresses limited to 80% of the

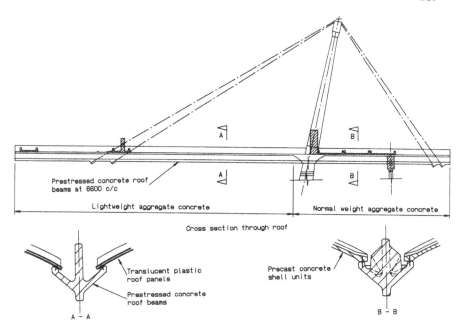

**Figure 7.15** Cross-section of the South Stand.

values allowed for normal weight concrete of the same grade, for a crack width up to 0.2 mm.

The stiffening beams also in lightweight concrete are provided at the cantilever end as well as at the points where the front row cables pick up the beam and at the grid line of the pylon. These beams have been post-tensioned longitudinally, integrating the whole roof primarily to control flutter and differential wind-induced movements of the beam [19].

Dynamic wind tunnel tests on the model of a similar structure showed that with such rigid longitudinal connections of the beams at cantilever ends and at the pylon's grids, a wind velocity over 55 m/s would be necessary to cause the flutter, whereas without these interconnections a wind speed as low as 18 m/s could initiate it. At the pylon head, an external damping device has been incorporated in the cable saddle box to reduce the oscillation of the beam due to vortex shedding.

*Grandstand for the Exhibition and Stampede at Calgary, Canada.* The complex, measuring 210 × 42.36 m on plan, is 34 m high and seats 17 350 people in three tiers; in addition, it accommodates seven floors of various utility areas (Figures 7.16 and 7.17). Opened in July 1974, this home for 'The Greatest Outdoor Show on Earth' is a useful functional building suitable for a whole range of activities from racing through to exhibition

**Figure 7.16** Exhibition and Stampede Grandstand at Calgary, showing post-tensioned lightweight concrete raker beams at upper tiers.

**Figure 7.17** General view of the Exhibition and Stampede Grandstand, Calgary.

and conferences. Large-scale prefabrication was used to enable completion of the entire construction in just 12 months, including five months of severe Canadian winter.

The roof construction, cantilevering 20 m, is in steel plate girders generally spaced at 6.1 m centres and supports profiled metal decking. The rest of the construction is in precast concrete consisting basically of H-frames and double-T units. H-frames are spaced at 12.2 m centres and 500 deep double-T's compositely with structural topping (minimum 50 mm) form the floor spanning between rows of frames. The interesting feature of the stand is the two upper tier rakers which are spaced at 12.2 m centres and cantilever 12 m supporting seating units. These rakers (upper and lower balcony units) are precast with lightweight concrete using 'expanded shale' aggregate and weigh about 40 tonnes each. Lightweight concrete was essential to reduce the size of the lifting crane as well as the total quantities of rebars and post-tensioning steel in the units. Each of the cantilevers was post-tensioned with six 7–18 mm strand tendons and two 12–12.7 mm strand tendons [20]. The concrete strength (cylinder) was 40 N/mm² (at 28 days) and density (air-dry) 1850 kg/m³.

*The West Stand of Newcastle United Football Club at St James Park, Newcastle, UK* (see Appendix 2).  The new West Stand of Newcastle United Football Club, covering a plan area of 130 × 32 m, provides 6300 seated accommodation and 39 private boxes, each accommodating 12 persons, in an environment of comfort and safety with a totally unobstructed view of the entire playing surface. In addition, the internal spaces of the stand in six levels have been utilised as 'concourse areas' and various facilities comprising sponsors' lounges, players' changing rooms, banqueting suite, club offices, ticket offices all with ancillary toilet accommodation. Based on column spacing of 9.2 m longitudinally and 5.5–10.6 m grid (as shown in Figures 7.18 and 7.19) transversely, the grandstand building provides a wide flexibility for internal planning and layout. The particular design was selected from several competitive tenders with 'design

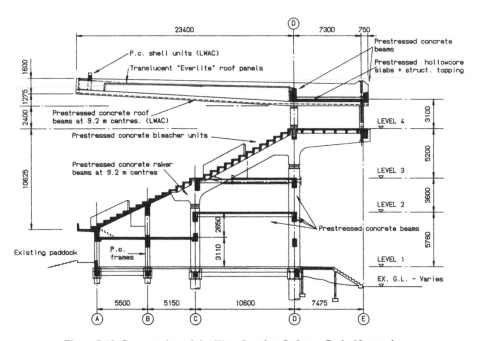

**Figure 7.18** Cross-section of the West Stand at St James Park, Newcastle.

**Figure 7.19** General view of the West Stand in use.

and build' submission, on considerations of safety, cost-effectiveness and the suitability of the proposal to meet the club's functional requirements.

The time constraint for the contract completion, and the need to keep the ongoing maintenance work to a minimum, led to the option of precast construction using a relatively small number of basic repetitive components that could be economically produced and easily assembled on site to reduce the site operation.

The special feature of the stand, no doubt, is the roof which spaced at 9.2 m centres, cantilevers 23.4 m supporting barrel-vaulted translucent PVC (Everlite) panels on the cantilever side and composite prestressed hollowcore slab at the rear side (i.e. over the private boxes). The heavier slab construction on the rear side increases the counterweight to the cantilever and also provides the required degree of insulation – thermal and sound – to the boxes.

The overall length of each beam is 31.5 m and the cross-section is an asymmetrical 'I' section (about the horizontal axis), the bottom flange being curved upwards to form built-in gutters for rainwater run-off. The depth of cross-section varies from 1600 to 2400 mm, as shown in Figure 7.17. the rear 8.1 m length (i.e. between supports) is in solid section to increase further the counterweight. The beams are cast in high-strength lightweight concrete using Lytag aggregate and each pretensioned with 26 15.2 mm diameter Dyform strands in the top flange, 8 12.6 mm diameter and 16 9.3 mm diameter low relaxation strands in the web and the bottom flange, some deflected to follow the profile of the soffit. The specified concrete strength of 70 N/mm², 28 days, has in fact been achieved at an average air-dry density of 1950 kg/m³. The concrete strength at release was 50 N/mm². Under uniaxial compression the stress–strain curve of lightweight concrete in general, and high-strength lightweight concrete in particular, shows reduced ductility indicating a tendency to brittle failure. To counteract this probability, transverse confining reinforcement has been provided in the compression zone for the entire length of these beams. It has been proved that even a nominal amount of confining reinforcement induces sufficient confining stresses, which in fact sets up a triaxial stress field in that zone, although the section may be under uniaxial loading [21]. In such conditions all concrete shows ductile property reducing the possibility of brittle failure.

The beams have positive supports on the front row of columns and are secured rigidly at the rear end (i.e. second line of support) to the raker frames below by vertical post-tensioning. The connection at the front row of the column is adequate for uplift when the beams are subjected to reversal of forces.

The direct benefit of using lightweight aggregate concrete is actually two-fold:

(1)  The 22% reduction in weight helped in handling, transport and erection of the unit and was reflected in the project cost.
(2)  As expected in this type of structure where the dead load/total load is high, substantial economy is achieved on material (concrete as well as steel). This also leads to saving in the frame and the foundation design.

The roof beams are designed as Class 3 structures (up to 0.2 mm crack width) in accordance with BS 8110 – the permissible tensile stress under the worst combination of service loads having been taken as 80% of the corresponding value given in the code for normal weight concrete element of the same depth, concrete strength and crack width.

At the cantilever end and at the support grids, stiffening beams have been provided which are post-tensioned longitudinally, interconnecting the roof beam rigidly in order to prevent wind-excited 'scissor oscillation' to which such light cantilever structures are susceptible.

## References

1. Bardhan-Roy, B. K., Design considerations for prestressed lightweight aggregate concrete, *The International Journal of Lightweight Concrete*, **2** (No. 4; December 1980).
2. *Design and Detailing of Concrete Structures for Fire Resistance*, Interim Guidance by a Joint Committee of the Institution of Structural Engineers and the Concrete Society, published by the Institution of Structural Engineers, April 1978.
3. *FIP Manual of Lightweight Aggregate Concrete*, 2nd edition, Surrey University Press, 1983.
4. Swamy, R. N., *Prestressed Lightweight Concrete* and *Development in Prestressed Concrete*, Applied Science Publications.
5. Holm, T. A., Performance of structural lightweight concrete in a marine environment, American Institute Publication SP-65, *Performance of Concrete in Marine Environment*.

6. Morgan, R., Development of the concrete hull, *Concrete Society Symposium on Ships and Floating Structures*, March 1977.
7. Gerwick Ben C., Jr, Current trends in design and construction of concrete ships, *Concrete Society Symposium on Ships and Floating Structures*, London, March 1977.
8. Morgan, R., Concrete for ship building, a challenge to LWC, *Proceedings of the 1st International Congress on Lightweight Concrete*, *London*, May 1968; Vol. 2, Discussion.
9. Berner, D, Gerwick Ben C., Jr and Polivka, M., Prestressed lightweight concrete in the transportation of cryogenic liquids, Paper presented at *Oceans 1983 Conference*.
10. Bamforth, P. B., Murray, W. T. and Browne, R. D., *The Application of Concrete Property Data at Cryogenic Temperature to LNG Tank Design*.
11. Standford, A. E., Concrete ship economics, *Concrete Society Symposium on Ships and Floating Structures*, *London*, March 1977.
12. Levi, F., Constantino, M, Perlazzene, R., Sonzogne, G. and Marioni, A., Prestressed concrete floating dry dock with a lifting capacity of 100 000, *FIP Symposium*, *Tiblisi*, 1972.
13. Short, A. and Kinniburg, W., *Lightweight Concrete*, 3rd edition, Applied Science Publishers, 1978.
14. Bobrowski, J., *Chairman's Report of the Commission on Lightweight Concrete Structures for the 7th Congress of the FIP*, New York, 1974.
15. Bomhard, H., Lightweight concrete structures, potentialities limits and realities, *The International Journal of Lightweight Concrete*, **2** (No. 4; December 1980).
16. Cover story, *New Civil Engineer*, 13 February 1986.
17. Wiig, M., Lightweight aggregates (LWAC) in high strength concrete: application in long span bridge construction, *FIP Notes*, 1992/4.
18. BS 8110: Part 1 and Part 2: 1985, Structural use of concrete, British Standards Institution, London.
19. Bobrowski, J., Abeler, P. W., Bardhan-Roy, B. K., The design of cantilever roofs to control dynamic effects of wind by external damping, *7th FIP Congress*, New York, 1974.
20. Bobrowski, J. and Bardhan-Roy, B. K., The application of prestressing in building, Paper presented at the *FIP Symposium*, Australia, 1976.
21. Clarke, J. L. and Pomeroy, C. D., Concrete opportunity for the structural engineer (a review of modified concretes for use in structures), *The Structural Engineer* (February 1985).

# Appendix 1

**Design of a road bridge using standard prestressed M beams in lightweight aggregate concrete**

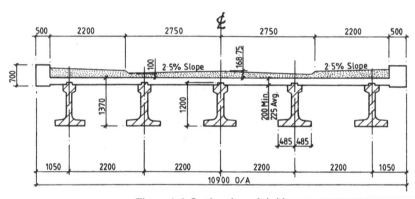

**Figure A.1** Section through bridge.

The design is based on a single-span two-lane side road overbridge crossing a typical by-pass around a medium sized market town. The road carried by the bridge is assumed to be minor and lightly trafficked.

The by-pass has dual 7.3 m wide carriageways and a 4.5 m central reservation. The span of the overbridge is 24.75 m between bearings and it consists of a precast prestressed M-beam in lightweight concrete carrying an *in-situ* reinforced concrete deck slab also in lightweight aggregate concrete. For both the precast and the *in-situ* concrete Lytag coarse aggregate is envisaged.

The substructure as well as the abutments and wing walls are assumed to be in normal weight concrete, and are not included in the design.

This exercise is a full and comprehensive analysis and design of the bridge deck and shows the application of BS 5400, Part 4, in the design of a prestressed bridge using lightweight aggregate concrete. Although the section on prestressed concrete in BS 5400, Part 4, does not actually cover lightweight aggregate concrete, it is nonetheless applicable with appropriate modification of the 'design values' given in the code for normal weight concrete design. The modification factors are based on experience in the use of prestressed lightweight aggregate concrete in buildings and incorporated in reference [4]. The exercise was originally carried out to supply the back-up information for the special publication of the Concrete Society on *Design and Cost Studies of Lightweight Concrete Highway Bridges* published in March 1986.

*Strength of concrete*

Precast  $f_{cu} = 50 \, \text{N/mm}^2$ (28 day)

$\qquad\quad f_{ci} = 40 \, \text{N/mm}^2$ (release)

Topping  $f_{cu} = 30 \, \text{N/mm}^2$ (28 day)

*E-values*

Precast  $E_{ci} = 31 \times \left(\dfrac{19}{23}\right)^2 = 21.155 \, \text{kN/mm}^2$ (release)

$\qquad\quad E_{c} = 34 \times \left(\dfrac{19}{23}\right)^2 = 23.2 \, \text{kN/mm}^2$ (28 day)

Topping  $E_{c} = 28 \times \left(\dfrac{18.5}{23}\right)^2 = 18.1 \, \text{kN/mm}^2$ (28 day)

*E*-values calculated in accordance with Clause 4.3.2.1 of BS 5400 Part 4.

*Allowable stresses in tension*

Design is based on Class 3(C), maximum design crack width being 0.25 mm.

Values in Table 25 have been multiplied by a factor 0.8 to allow for lightweight aggregate concrete in accordance with references [4] and [5].

*Shrinkage, creep and thermal expansion co-efficient*

Shrinkage strain and specific creep for lightweight aggregate concrete has been taken as $300 \times 10^{-6}$ and $72 \times 10^{-6}$ per unit stress (N/mm$^2$) respectively on the basis of manufacturers' recommendations as well as suggestions in references [4–6].

The thermal expansion co-efficient has been taken as $7 \times 10^{-6}$ per °C.

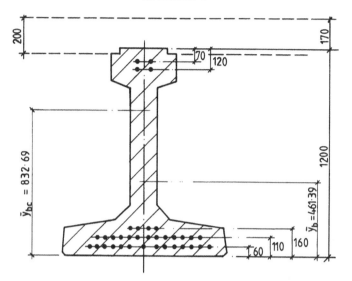

**Figure A.2** Properties of the beam.

*Precast section*

$A = 396\,675\,\text{mm}^2$

$I = 66.789 \times 10^9\,\text{mm}^4$

$Z_t = 90.425 \times 10^6\,\text{mm}^3$

$Z_b = 144.756 \times 10^6\,\text{mm}^3$

$\bar{y}_b = 461.39\,\text{mm}$

$i^2 = \dfrac{I}{A} = 168\,372\,\text{mm}^2$

*Composite section*

$m = \dfrac{E_{in\text{-}situ}}{E_{\text{precast}}} = \dfrac{18.1}{23.2} = 0.78$

$A = 731\,919\,\text{mm}^2$

$I_c = 187.266 \times 10^9\,\text{mm}^4$

$Z_{t(\text{top of }in\text{-}situ)} = 446.826 \times 10^6\,\text{mm}^3$ (in terms of weaker concrete)

$Z_{tpc} = 509.830 \times 10^6\,\text{mm}^3$

$Z_{bc} = 224.892 \times 10^6\,\text{mm}^3$

$\bar{y}_{bc} = 832.69\,\text{mm}$

*Prestressing force and its CG*

Area of prestressing steel    $= 4 \times 94.2 + 33 \times 100.5 = 3693.3 \, \text{mm}^2$

$P_i$ = initial prestressing force $= (4 \times 165) \times 0.7 + (33 \times 184) \times 0.7$

$\qquad\qquad\qquad\qquad\qquad = 4712.40 \, \text{kN}$

CG of prestressing force    $= 192.5 \, \text{mm from bottom}$

$\therefore e$ = eccentricity $= 461.39 - 192.5 = 268.89 \, \text{mm}$

$P_i$ = Average stress (initial) in steel $= \dfrac{4712.40 \times 10^3}{3693.30} = 1276 \, \text{N/mm}^2$

Initial stresses in concrete (solely due to prestress)

$f_{ti}$ (top of precast)    $= \dfrac{4712.4 \times 10^3}{396675} - \dfrac{4712.40 \times 10^3 \times 268.89}{90.425 \times 10^6}$

Stress due to o/w    $= (-)2.13 \, \text{N/mm}^2 \, (\text{T})^*$

$f_{bi}$ (bottom of precast) $= \dfrac{4712.40 \times 10^3}{396675} + \dfrac{4712.40 \times 10^3 \times 268.8}{144.756 \times 10^6}$

$\qquad\qquad\qquad\qquad = 20.63 \, \text{N/mm}^2 \, (\text{C})^*$

$f_i$ (at CG of steel)    $= \dfrac{(20.63 + 2.13)}{1200} \times (1200 - 192.5) - 2.13$

$\qquad\qquad\qquad\qquad = 16.98 \, \text{N/mm}^2 \, (\text{C})$

Let $x$ = % instantaneous loss due to elastic shortening

$\therefore \quad x \quad = \dfrac{m_1 K \times 100}{(1 + m_1 K)}$  (see reference [5])

$E_{steel} = 195.00 \, \text{kN/mm}^2$

$m_1 \quad = \dfrac{E_{steel}}{E_{ci}} = \dfrac{195.00}{21.155} = 9.2$

$K \quad = \dfrac{A_{st}}{A} \times \left(1 + \dfrac{e^2}{i^2}\right) = \dfrac{3693.30}{396675} \times \left(1 + \dfrac{268.89^2}{168372}\right)$

$\qquad\qquad\qquad = 0.0133$

$\therefore x \quad = \dfrac{9.20 \times 0.0133 \times 100}{(1 + 9.2 \times 0.0133)} = 10.9\%$

Initial stresses in concrete after instantaneous loss

$f_{ti} = (-)2.13 \times 0.899 = (-)1.91 \, \text{N/mm}^2 \, (\text{T})$

---

*T = tension, C = compression.

with co-existant self-weight. This will be in compression hence OK. At either end 2 Nos strands at third row (from bottom) will be debonded for a length of 1.5 m.

$$f_{bi} = 20.36 \times 0.899 = 18.30\,\text{N/mm}^2 < 0.5\,f_{ci} \quad \therefore \text{OK}$$
$$\text{(comp.)}$$
$$f_i = 16.98 \times 0.899 = 15.27\,\text{N/mm}^2 \,\text{(C)}$$

Stress in concrete at CG of steel due to initial prestress (after elastic loss) and self-weight bending moment $= 15.27 - 2.33 = 12.94\,\text{N/mm}^2$ (C)

*Losses*

$$\text{N/mm}^2$$

1. Relaxation @ 2.5% of initial stress $= 0.025 \times 1276 \quad = \;\; 31.90$
2. Shrinkage $= 300 \times 10^{-6} \times 195 \times 10^3 \qquad\qquad = \;\; 58.50$
3. Creep $\;\; = 72 \times 10^{-6} \times 0.85 \times 195 \times 10^3 \times 12.94 \;\; = 154.80$
4. Elastic shortening $= \dfrac{10.9}{100} \times 1276 \qquad\qquad\quad = 139.00$

$$\Sigma = 384.20$$

$$\% \;\text{loss} = \frac{384.20}{1276} \times 100 = 30.11\%$$

Final stresses (solely due to prestress) after all losses:

$$f_t = 0.699 \times 2.13 = (-)1.49\,\text{N/mm}^2 \,\text{(T)}$$
$$f_b = 0.699 \times 20.63 = (+)14.42\,\text{N/mm}^2 \,\text{(C)}$$

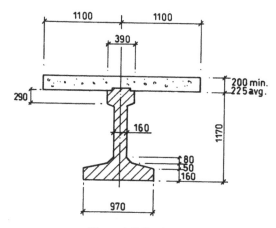

**Figure A.3** Section.

*Loading (per beam)*

(a) *SW of PC* $= 19 \times 396\,675 \times 10^{-6} = 7.537\,\text{kN/m}$

$$\text{SW bending moment (max.) per beam} = \frac{7.537 \times 24.75^2}{8}$$
$$= 577.11\,\text{kNm}$$

(b) *Topping* (av. 225 mm) $= 18.5 \times 0.225 \times 9.9/5$
$$= 8.242\,\text{kN/m per beam}$$

(c) *Edge beam* $= 18.5 \times 0.5 \times 0.7 \times 2/5 = 2.59\,\text{kN/m per beam}$

Bending moment due to topping and edge beam

$$= \frac{(8.24 + 2.59) \times 24.75^2}{8} = 829.26\,\text{kNm/per beam}$$

(d) *Finishes (asphalt)*: total cross section area $= 1.74\,\text{m}^2$
$$= 1.74 \times 23/5 \qquad = 8.00\,\text{kN/m}$$

$$\text{Bending moment} \quad = \frac{8.0 \times 24.75^2}{8} \qquad = 612.56\,\text{kNm}$$

(e) *Live load*

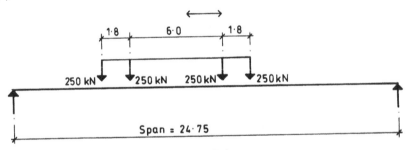

**Figure A.4**

  (i) HB loading (Figure A.4)
    For most critical position of HB loading see Figure A.6
 (ii) HA loading

    UDL             $= 30\,\text{kN/m per notional lane}$
    Knife edge load $= 120\,\text{kN per notional lane}$

*Longitudinal bending due to HB loads*

Transverse position of wheels for most critical effect (as shown in Figure A.5)

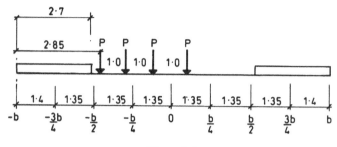

**Figure A.5**

$\lambda_p$ at    $-b/2$    $-b/4$    $0$    $b/4$

$$1.11p \quad 1.33p \quad 1.26p \quad 0.3p$$

These values are calculated by assuming a beam of 1.35 m simply supported at successive reference stations.

*Position of wheels along span*

For maximum bending moment: see Figure A.6.
For maximum vertical shear: see Figure A.6a.

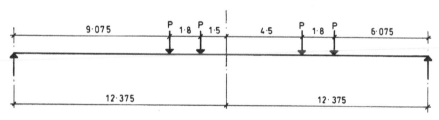

**Figure A.6**

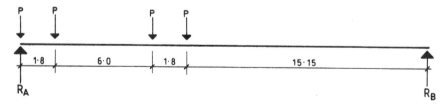

**Figure A.6a**

$P = 250\,\text{kN}$ (25 units HB loading), total width $= 2b = 10.9\,\text{m}$

*Maximum bending moment:*

$R_A = 439.4\,\text{kN}, \quad R_B = 560.6\,\text{kN}$

$M_{\text{max.}}$ at centre (for full width) $= 439.4 \times 12.375 - 250(1.5 + 3.3)$
$$= 4237.575\,\text{kNm}$$

$M_{\text{max.}}$ per m width $= 388.77\,\text{kNm/m}$ width

$M_{\text{max.}}$ per beam (av.) $= 388.77 \times 2.2 = \underline{855.29\,\text{kNm}}$

*Notation*

$2a \quad = 2 \times 12.375 = 24.75\,\text{m}$ (span)

$2b \quad = 2 \times 5.45 \quad = 10.90\,\text{m}$ (width)

$c \quad =$ Distance along span of load

$D_x \quad = Ei$

$i \quad = I/P$

$P \quad =$ Spacing of beams $= 2.200\,\text{m}$

$D_y \quad = Ej$

$j \quad =$ Transverse $2^{\text{nd}}$ moment of area $= J/q$

$J \quad =$ Transverse $2^{\text{nd}}$ moment of area of stiffener

$q \quad =$ Spacing of stiffeners

$D_{xy} =$ Longitudinal torsional rigidity $= Gi_0$

$i_0 \quad = \Sigma\gamma bd^3$

$D_{yx} =$ Transverse torsional rigidity $= Gj_0$

$j_0 \quad = \Sigma\gamma ld^3 =$ torsional rigidity per unit length

$D_1 \quad =$ Coupling rigidity per unit width $= UD_{xy} = UEi_0$

$D_2 \quad =$ Coupling rigidity per unit span $= UD_{yx} = UEj_0$

$2H \quad =$ Total torsional rigidity $= 2a\sqrt{D_xD_y}$

$\theta \quad =$ Flexural parameter $= \dfrac{b}{L}\sqrt[4]{\dfrac{D_x}{D_y}}$

$L \quad =$ Span of deck $= 2a$

$\alpha \quad = \dfrac{i_0 + j_0}{\sqrt{(ij)}} \times \dfrac{1}{4}$

*For this particular case*

$2a \quad = 24.75\,\text{m}$

$2b \quad = 10.90\,\text{m}$

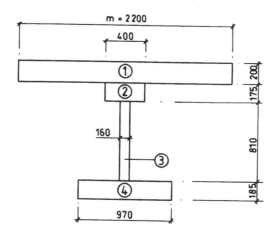

**Figure A.7** Idealised section for flexural parameter.

$P = 2.20\,\text{m}$

$i = I/P = \dfrac{187.266 \times 10^9}{2.2} = 85.121 \times 10^9\,\text{m}$

$E = 23.202\,\text{kN/mm}^2$ (PC beam only)

$D_x = Ei = 23.202 \times 10^3 \times 85.121 \times 10^9 = 1.975 \times 10^{15}\,\text{Nmm}^2/\text{m}$

$j = J \text{ per } m = \dfrac{1000 \times 200^3}{12} \times 0.78 = 0.52 \times 10^9\,\text{Nmm}^4/\text{m}$

$D_y = Ej = 23.202 \times 10^3 \times 0.52 \times 10^9 = 1.207 \times 10^{13}\,\text{Nmm}^2/\text{m}$

$\theta = \dfrac{b}{L}\sqrt[4]{\dfrac{D_x}{D_y}} = \dfrac{5450}{24\,750} \times \sqrt[4]{\dfrac{1.975 \times 10^{15}}{1.207 \times 10^{13}}} = 0.788$

$D_{xy} = Gi_0$

$i_0 = (1)\ \dfrac{b}{a} = 8.58, \quad \gamma = 0.306 \quad \begin{bmatrix} 2a = 200 \\ 2b = 0.87 \times 2200 = 1716 \end{bmatrix}$

(Refer to Figure A.7)

$(2)\ \dfrac{b}{a} = 2.28, \quad \gamma = 0.241 \quad \begin{bmatrix} 2a = 175 \\ 2b = 400 \end{bmatrix}$

$(3)\ \dfrac{b}{a} = 5.0625,\ \gamma = 0.291 \quad \begin{bmatrix} 2a = 160 \\ 2b = 810 \end{bmatrix}$

$(4)\ \dfrac{b}{a} = 5.243, \quad \gamma = 0.292 \quad \begin{bmatrix} 2a = 185 \\ 2b = 970 \end{bmatrix}$

$i_0 = 0.306 \times 200^3 \times 1716/2 + 0.241 \times 175^3 \times 400 + 0.291 \times 160^3$
$\times 810 + 0.292 \times 185^3 \times 970$

$= \underline{5.38 \times 10^9}$

$$j_o = 0.78 \times 1000 \times 200^3/6 \qquad = \underline{1.04 \times 10^9}$$

$$\alpha = \frac{(5.38 + 1.04) \times 10^9}{\sqrt{(85.121 + 0.52) \times 10^9}} \times \frac{1}{4} = \underline{0.173}$$

$$\sqrt{\alpha} \qquad\qquad\qquad = \underline{0.416}$$

*Distribution factors*

$\theta = 0.788$, $\alpha = 0.173$, $\sqrt{\alpha} = 0.416$

|  | | Reference station | | | | | | | | | |
|---|---|---|---|---|---|---|---|---|---|---|---|
| Load at | $-b$ | $-3b/4$ | $-b/2$ | $-b/4$ | $0$ | $b/4$ | $b/2$ | $3b/4$ | $b$ | $\Sigma/L$ |
| $K_0$ (for $\alpha = 0$) | $-b/2$ | $-0.48$ | $-0.19$ | $0.15$ | $0.55$ | $1.04$ | $1.62$ | $2.05$ | $2.02$ | $1.86$ | $0.994$ |
| | $-b/4$ | $-0.48$ | $0.02$ | $0.55$ | $1.10$ | $1.63$ | $1.86$ | $1.63$ | $1.10$ | $0.43$ | $0.995$ |
| | $0$ | $-0.29$ | $0.40$ | $1.02$ | $1.66$ | $1.86$ | $1.66$ | $1.02$ | $0.40$ | $-0.29$ | $0.988$ |
| | $+b/4$ | $0.43$ | $1.10$ | $1.63$ | $1.86$ | $1.63$ | $1.10$ | $0.55$ | $0.02$ | $-0.48$ | $0.995$ |
| $K_1$ (for $\alpha = 1$) | $-b/2$ | $0.26$ | $0.35$ | $0.47$ | $0.68$ | $0.98$ | $1.33$ | $1.61$ | $1.63$ | $1.55$ | $0.995$ |
| | $-b/4$ | $0.41$ | $0.52$ | $0.68$ | $0.93$ | $1.21$ | $1.42$ | $1.33$ | $1.14$ | $0.99$ | $0.995$ |
| | $0$ | $0.62$ | $0.78$ | $0.98$ | $1.22$ | $1.38$ | $1.22$ | $0.98$ | $0.78$ | $0.62$ | $0.997$ |
| | $+b/4$ | $0.99$ | $1.14$ | $1.33$ | $1.42$ | $1.21$ | $0.93$ | $0.68$ | $0.52$ | $0.41$ | $0.995$ |

*Derivation of distribution coefficient profile*

| Load at | $\lambda$ | $-b$ | $-3b/4$ | $-b/2$ | $-b/4$ | $0$ | $b/4$ | $b/2$ | $3b/4$ | $b$ |
|---|---|---|---|---|---|---|---|---|---|---|
| $-b/2$ | $1.11$ | $-0.53$ | $-0.21$ | $0.17$ | $0.61$ | $1.15$ | $1.80$ | $2.28$ | $2.24$ | $2.06$ |
| $-b/4$ | $1.33$ | $-0.64$ | $0.03$ | $0.73$ | $1.46$ | $2.17$ | $2.47$ | $2.17$ | $1.46$ | $0.57$ |
| $0$ | $1.26$ | $-0.37$ | $0.50$ | $1.29$ | $2.09$ | $2.34$ | $2.09$ | $1.29$ | $0.50$ | $-0.37$ |
| $b/4$ | $0.3$ | $0.13$ | $0.33$ | $0.49$ | $0.56$ | $0.49$ | $0.33$ | $0.17$ | $0.01$ | $-0.14$ |
| $\Sigma\lambda K_0$ | | $-1.41$ | $0.65$ | $2.68$ | $4.72$ | $6.15$ | $6.69$ | $5.91$ | $4.21$ | $2.12$ |
| $\Sigma\lambda K_0/4$ | | $-0.35$ | $0.16$ | $0.67$ | $1.18$ | $1.54$ | $1.67$ | $1.48$ | $1.05$ | $0.53$ |
| $-b/2$ | $1.11$ | $0.29$ | $0.39$ | $0.52$ | $0.75$ | $1.09$ | $1.48$ | $1.79$ | $1.81$ | $1.72$ |
| $-b/4$ | $1.33$ | $0.56$ | $0.69$ | $0.90$ | $1.24$ | $1.61$ | $1.89$ | $1.77$ | $1.52$ | $1.32$ |
| $0$ | $1.26$ | $0.78$ | $0.98$ | $1.23$ | $1.54$ | $1.74$ | $1.54$ | $1.23$ | $0.98$ | $0.78$ |
| $+b/4$ | $0.30$ | $0.30$ | $0.34$ | $0.40$ | $0.43$ | $0.36$ | $0.28$ | $0.20$ | $0.16$ | $0.12$ |
| $\Sigma\lambda K_1$ | | $1.92$ | $2.40$ | $3.05$ | $3.96$ | $4.80$ | $5.19$ | $4.99$ | $4.47$ | $3.94$ |
| $\Sigma\lambda K_1/4$ | | $0.48$ | $0.60$ | $0.76$ | $0.99$ | $1.20$ | $1.30$ | $1.25$ | $1.12$ | $0.99$ |
| $\Sigma\lambda K_1/4 - \Sigma\lambda K_0/4$ | | $0.83$ | $0.44$ | $0.09$ | $-0.19$ | $-0.34$ | $-0.37$ | $-0.23$ | $0.07$ | $0.46$ |
| $(\Sigma\lambda K_1/4 - \Sigma\lambda K_0/4 \times \sqrt{\alpha}$ | | $0.35$ | $0.18$ | $0.04$ | $-0.08$ | $-0.14$ | $-0.15$ | $-0.10$ | $0.03$ | $0.19$ |
| $K_\alpha = (\Sigma\lambda K_1/4 - \Sigma\lambda K_0/4)$ $\times \sqrt{\alpha} + \Sigma\lambda K_0/4$ | | $0.00$ | $0.34$ | $0.71$ | $1.10$ | $1.40$ | $1.52$ | $1.38$ | $1.08$ | $0.72$ |

*Notes*

1. $\Sigma/L$ Column based on Simpson's Rule

$$\frac{1}{3}\frac{L}{8} \times \frac{1}{L}[a_0 + 4a_1 + 2a_2 + 4a_3 + 2a_4 + 4a_5 + 2a_6 + 4a_7 + a_8]$$

2. Reference stations are reversed from Rowe's book
   i.e. reference station $-b/4$ is $+b/4$ in Rowe.

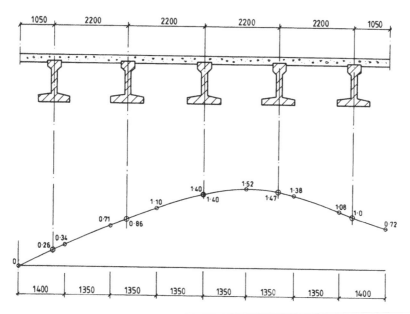

**Figure A.8**

Maximum $K_\alpha = 1.47$ at any beam
Maximum $M_{LL} = 855.29 \times 1.47 = 1257.28\,kNm$

*Bending moment due to HA loading*

Span      $= 24.75\,m$
HA load $= 30\,kN/m/notional\ lane$
          $= 30 \times 2 = 60\,kN/m\ total$
KEL      $= 120\,kN/notional\ lane$
          $= 120 \times 2 = 240\,kN\ total$
Total $M = \dfrac{60 \times 24.75^2}{8} + 240 \times \dfrac{24.75}{4} = 6079\,kNm = 1220\,kNm/beam$

*HB loading*

Effect is more severe (see pages 202 and 205) and taken for design.

Maximum stresses at mid-span due to dead and live load and prestress

A  [self-weight + prestress]: (precast section only)

$$f_{tpc} = \frac{577.11 \times 10^6}{90.425 \times 10^6} - 1.49 = 4.89 \, \text{N/mm}^2 \, (C)$$

$$f_b = -\frac{577.11 \times 10^6}{144.756 \times 10^6} + 14.42 = 10.43 \, \text{N/mm}^2 \, (C)$$

B  [topping + edge beam]: (precast section only)

$$f_{tpc} = \frac{829.26 \times 10^6}{90.425 \times 10^6} = 9.17 \, \text{N/mm}^2 \, (C)$$

$$f_b = -\frac{829.26 \times 10^6}{144.756 \times 10^6} = (-)5.73 \, \text{N/mm}^2 \, (T)$$

C  [finishes, i.e. asphalt]: (composite section)

$$f_{tpc} \, (\text{top of screed}) = \frac{612.56 \times 10^6}{509.83 \times 10^6} \times 1.2^* = 1.44 \, \text{N/mm}^2 \, (C)$$

$$f_t \, (\text{top of } in\text{-}situ) = \frac{612.56 \times 10^6}{446.826 \times 10^6} \times 1.2^* = 1.64 \, \text{N/mm}^2 \, (C)$$

$$f_b = (-)\frac{612.56 \times 10^6}{224.892 \times 10^6} \times 1.2^* = (-)3.26 \, \text{N/mm}^2 \, (T)$$

  *Factor for superimposed DL for SLS design (BS 5400, Part 2).

D  [Live load]: (composite section)

$$f_t \, (\text{top of } in\text{-}situ) = \frac{1257.28 \times 10^6}{446.826 \times 10^6} = 2.81 \, \text{N/mm}^2 \, (C)$$

$$f_{tpc} \, (\text{top of precast}) = \frac{1257.28 \times 10^6}{509.83 \times 10^6} = 2.47 \, \text{N/mm}^2 \, (C)$$

$$f_b = (-)\frac{1257.28 \times 10^6}{224.892 \times 10^6} = (-)5.59 \, \text{N/mm}^2 \, (T)$$

*Thermal stresses*

BS 5400 Part 2 (Figure 9 Group 4)

1. *Temperature distribution* (as shown in Figures A.9 and A.10)

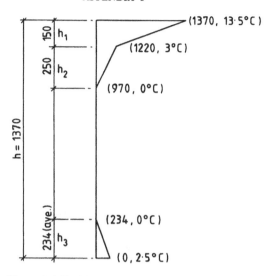

**Figure A.9** Positive temperature difference.

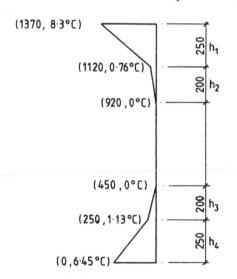

**Figure A.10** Reverse temperature difference.

Topping $\quad f_{cu} = 30\,\text{N/mm}^2 \quad E = 19.11\,\text{kN/mm}^2$

Beam $\quad\quad f_{cu} = 50\,\text{N/mm}^2 \quad E = 23.20\,\text{kN/mm}^2$

$\quad\quad\quad\quad \alpha = 7 \times 10^{-6}/°\text{C}$

In topping $\quad f = 7 \times 10^{-6} \times 19.11 \times 10^3 = 0.1338\,\text{N/mm}^2/°\text{C}$

In beam $\quad\quad f = 7 \times 10^{-6} \times 23.20 \times 10^3 = 0.1624\,\text{N/mm}^2/°\text{C}$

*Applied stress distribution*

**Figure A.11**                          **Figure A.12**

*CG of forces positive temperature* (Figures A.9 and A.11)

$$2200 \times 150 \times \frac{1}{2}(0.4 + 1.81) \qquad = 364.65\,\text{kN} \times \left(1220 + \frac{150}{3} \times \frac{4.02}{2.21}\right)$$

$$= 478\,038\,\text{kNmm}$$

$$2200 \times 35 \times \frac{1}{2} \times 0.89 \qquad = 34.27\,\text{kN} \times \left(1185 + \frac{35}{3} \times \frac{1.29}{0.89}\right)$$

$$= 41\,195\,\text{kNmm}$$

$$390 \times 155 \times \frac{1}{2} \times 0.63 \qquad = 19.042\,\text{kN} \times \left(1030 + \frac{155}{3} \times \frac{1.12}{0.63}\right)$$

$$= 21\,362\,\text{kNmm}$$

$$\frac{1}{2} \times 0.14 \times 60 \times \frac{1}{2} \times 550 \qquad = 1.155\,\text{kN} \times \left(970 + \frac{60}{3} \times \frac{940}{550}\right)$$

$$= 1\,160\,\text{kNmm}$$

$$970 \times 160 \times \frac{1}{2} \times 0.54 \qquad = 41.904\,\text{kN} \times \left(\frac{160}{3} \times \frac{0.67}{0.54}\right)$$

$$= 2\,773\,\text{kNmm}$$

$$\frac{1}{2} \times 0.13 \times \frac{1}{2} \times (350 + 970) \times 74 = 3.175\,\text{kN} \times \left(160 + \frac{50}{3} \times \frac{1670}{1320}\right)$$

$$= \underline{575\,\text{kNmm}}$$

$$\Sigma = \overline{464.2\,\text{kN}} \qquad \Sigma = 545\,103\,\text{kNmm}$$

$$\text{CG of temperature force} = \frac{545\,103}{464.2} = 1174.4\,\text{mm from the bottom}$$

$$e \qquad = 1174.4 - 832.69 = 341.71\,\text{mm}$$

$$m \qquad = 464.2 \times 0.342 = 158.76\,\text{kNm}$$

$$f_{t(in\text{-}situ)} = -\left(\frac{464.2 \times 10^3}{731\,919} + \frac{158.76 \times 10^6}{446.82 \times 10^6}\right) \times 0.8^* = -0.79\,\text{N/mm}^2\,\text{(T)}$$

$$f_{tpc\,(top\,of\,precast)} = \left(-0.634 - \frac{158.76}{509.83}\right) \times 0.8^* = -0.75\,\text{N/mm}^2\,\text{(T)}$$

$$f_b = \left(-0.634 + \frac{158.76}{224.892}\right) \times 0.8^* = 0.06\,\text{N/mm}^2\,\text{(C)}$$

*CG of forces reverse temperature* (Figures A.10 and A.12)

| | | | | | |
|---|---|---|---|---|---|
| $2200 \times 185 \times \frac{1}{2} \times 1.49$ | = | 303.2 | kN $\times$ 1292.61 = | 391\,919 kNm |
| $390 \times 155 \times \frac{1}{2} \times 0.45$ | = | 13.6 | kN $\times$ 1125.3 = | 15\,305 kNm |
| $\frac{1}{2} \times 550 \times 60 \times \frac{1}{2} \times 0.1$ | = | 0.825 kN $\times$ 1004 | = | 828 kNm |
| $160 \times 50 \times 0.015$ | = | 0.12 | kN $\times$ 953.3 = | 114 kNm |
| $970 \times 160 \times \frac{1}{2} \times 1.54$ | = | 119.5 | kN $\times$ 70.3 = | 8\,401 kNm |
| $\frac{1}{2} \times 1225 \times 90 \times \frac{1}{2} \times 0.67$ = | | 18.5 | kN $\times$ 196.2 = | 3\,624 kNm |
| $160 \times 200 \times \frac{1}{2} \times 0.18$ | = | 2.88 | kN $\times$ 350 = | 1\,008 kNm |
| | | $\Sigma = \overline{458.6\quad\text{kN}}$ | | $\Sigma = \overline{421\,199\,\text{kNm}}$ |

$$\text{CG of temperature force (reverse)} = \frac{421\,199}{458.6} = 918.4\,\text{mm}$$

$e = 918.4 - 832.69 = 85.71\,\text{mm}$

$M = 0.08571 \times 458.6 = 39.31\,\text{kNm}$

$$f_t = \left[\frac{458.6 \times 10^3}{731919} + \frac{39.31}{446.82}\right] \times 0.8^* = 0.57\,\text{N/mm}^2\,\text{(C)}$$

$$f_{tpc} = \left[0.627 + \frac{39.31}{509.83}\right] \times 0.8^* = 0.56\,\text{N/mm}^2\,\text{(C)}$$

$$f_b = \left[0.627 - \frac{39.31}{224.892}\right] \times 0.8^* = 0.36\,\text{N/mm}^2\,\text{(C)}$$

*Factor for SLS design (BS 5400, Part 2).

*Summary of prestress* (critical combination)

| Point | Prestress + SW | Topping and edge beam | Finish | *L.L* | Thermal | Total |
|---|---|---|---|---|---|---|
| Top of *in-situ* | | | 1.64 (C) | 2.81 (C) | 0.57 (C) | 5.02 (C) |
| Top PC | 4.89 (C) | 9.17 (C) | 1.44 (C) | 2.47 (C) | 0.56 (C) | 18.53 (C) |
| Bottom fibre | 10.43 (C) | −5.73 (T) | −3.26 (T) | −5.59 (T) | 0.05 (C) | −4.1 (T) |

*Note*: Under no live load bottom fibre is in compression Class 3 allowable.

Allowable compression $= 0.33 \times 30 = 9.9 \, \text{N/mm}^2$ topping

$$= 0.4 \times 50 = 20.0 \, \text{N/mm}^2 \text{ top PC}$$

Allowable tension $= -7.8 \times 0.7 \times 0.8 = -4.4 \, \text{N/mm}^2$
Class 3 (0.25 crack width)

Section is OK

*Check for ultimate*

$M_u = [1.15 \times (577.11 + 829.26) + 1.75 \times 612.56 + 1.1 \times 1257.28 + 1.0 \times 158.76] \times 1.15$

$\quad = 4865 \, \text{kNm}$

$$\frac{f_{pu} \, Aps}{f_{cu} \, bd} = \frac{29 \times 184 \times 10^3}{30 \times 2200 \times (1370 - 84^*)} = 0.063$$

$$\frac{f_{pd}}{0.87 \times f_{pu}} = 1.0 \text{ (See reference [1], table 27)}$$

*Consider only strands in bottom two rows*

$F_{ult} = 29 \times 0.87 \times 184 = 4642.3 \, \text{kN}$

$f_{cu} = 30 \, \text{N/mm}^2$ topping $\qquad f_c' = 0.4 \times 30 = 12 \, \text{N/mm}^2$

$$d_c = \frac{4642.3 \times 10^3}{2200 \times 0.4 \times 30} = 176$$

$l_A = 1370 - 88 - 84^* = 1198 \, \text{mm}$

$M_R = 1.198 \times 4642.3 = \underline{5561 \, \text{kNm}} > M_u$

* CG of strands

*Maximum ultimate vertical shear*

1. Shear (ch) due to DL excluding finishes

$$= (7.537 + 8.242 + 2.59) \times \frac{24.75}{2} = 227.32 \, \text{kN/beam}$$

2. Shear (ch) due to finishes $= 8 \times \frac{24.75}{2} = 99.0 \, \text{kN/beam}$

3. Shear (ch) due to live load (see Figure A.6a)

$$R_A = 4 \times 250 \times \frac{19.95}{24.75} = 806.06 \, \text{kN}$$

$$R_B = 1000 - 806.06 = 193.94 \, \text{kN}$$

$\therefore$ Shear at A $= 806.06 \, \text{kN/5 beams}$

Assuming distribution factor of 1.47 (same as for bending moment):

Shear at A $= \dfrac{806.06}{5} \times 1.47 = 237.0 \, \text{kN}$

Ultimate shear $= 1.15 \times 227.32 + 1.75 \times 99 + 1.3 \times 237$
$$= 261.42 + 173.25 + 308.1$$
$$= 742.77 \, \text{kN}$$

Depth of prestressed beam $= 1170 \, \text{mm}$

Depth of composite beam $= 1370 \, \text{mm}$

2 no. 12.9 $\phi$ supastrand in third layer from bottom debonded.
CG of $15 + 14 + 2$ i.e. 31 no. 12.9 $\phi$ supastrand at bottom and 4 no. 12.5 diameter strands at top $= 198.65 \, \text{mm}$ from bottom.

Initial prestressing force $= 4 \times 165 \times 0.7 + 31 \times 184 \times 0.7$
$$= 462 + 3992.8$$
$$= 4454.8 \, \text{kN}$$

$f_{ti}$ (initial stress at top of beam) $= -1.714 \, \text{N/mm}^2$ (T)

$f_{bi}$ (initial stress at bottom of beam) $= 19.316 \, \text{N/mm}^2$ (C)

$Y_{bot}$ (distance of centroidal axis from bottom) $= 832.69 \, \text{mm}$

Initial prestress at centroidal axis $= 4.723 \, \text{N/mm}^2$

After allowing for 30% loss, prestress $= 0.7 \times 4.723$
$$= 3.306 \, \text{N/mm}^2$$

Transmission length $l_t = \dfrac{K_t \phi}{\sqrt{f_{ci}}} = \dfrac{240 \times 12.9}{\sqrt{40}} = 490 \, \text{mm}$

% of prestress developed at 300 mm from end of beam

$= \dfrac{300}{490}$, i.e. 61.2%

$\therefore f_{cp} = 0.612 \times 3.306 = 2.023 \, \text{N/mm}^2$

*Section uncracked in flexure*

Ultimate shear resistance $(V_{co}) = 0.67 \, bh \, \sqrt{f_t^2 + f_{cp} f_t}$

$f_t = 0.24 \, \sqrt{f_{cu}} = 0.24 \, \sqrt{50} = 1.697 \, \text{N/mm}^2$

$V_{co} = 0.67 \times 160 \times 1370 \, \sqrt{1.697^2 + 2.023 \times 1.697}$

$\quad = 369.00 \times 10^3 \, \text{N} < 742.77 \times 10^3 \, \text{N}$

$\therefore$ Shear reinforcement required

$$\frac{A_{sv}}{S_v} = \frac{V + 0.4 \, bd_t - V_{co}}{0.87 f_{yv} \, dt}$$

$$= \frac{742770 + 0.4 \times 160 \times 1285 - 369000}{0.87 \times 460 \times 1285}$$

$$= 0.89$$

$\therefore S_v = \dfrac{A_{sv}}{0.89} = \dfrac{157}{0.89} = 176 \, \text{mm}$

where $A_{sv}$ (2T10 links) $= 157 \, \text{mm}^2$ provide T10 links at 150 crs

*Transverse bending moments*

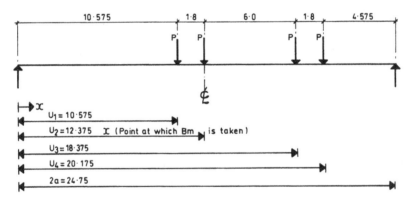

**Figure A.13** Longitudinal position of axles of maximum transverse moment.

*Transverse position of wheels for maximum transverse moment*

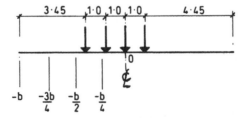

**Figure A.14**

$\theta = 0.788$

$2\theta = 1.576$

$3\theta = 2.364$

$4\theta = 3.152$

$\alpha = 0.173$

Values of $\mu_0$ at reference station 0 ($\times 10^4$)

| Values at $\theta$ | Load at | | | | | Wheel positions* | | | | |
|---|---|---|---|---|---|---|---|---|---|---|
| | 0 | $b/4$ | $b/2$ | $3b/4$ | $b$ | 1 | 2 | 3 | 4 | $\Sigma\mu_0$ |
| 0.788 | 1540 | 510 | −140 | −580 | −960 | 180 | 750 | 1540 | 750 | 3220 |
| 1.576 | 710 | −40 | −150 | −75 | 20 | −140 | 100 | 710 | 100 | 770 |
| 2.364 | 500 | −100 | −40 | 0 | 10 | −80 | −75 | 500 | −75 | 270 |
| 3.152 | 375 | −100 | 0 | 0 | 0 | −55 | −85 | 375 | −85 | 150 |

* See influence line curves (Figure A.18).

Values of $\mu_1$ at reference station 0 ($\times 10^4$)

| Values at $\theta$ | Load at | | | | | Wheel positions* | | | | |
|---|---|---|---|---|---|---|---|---|---|---|
| | 0 | $b/4$ | $b/2$ | $3b/4$ | $b$ | 1 | 2 | 3 | 4 | $\Sigma\mu_1$ |
| 0.788 | 1140 | 300 | −100 | −160 | −235 | 100 | 480 | 1140 | 480 | 2200 |
| 1.576 | 580 | 15 | −40 | −30 | −20 | −15 | 80 | 580 | 80 | 725 |
| 2.364 | 380 | 0 | −30 | 0 | 0 | −20 | 50 | 380 | 50 | 460 |
| 3.152 | 295 | 0 | −30 | 0 | 0 | −20 | 50 | 295 | 50 | 375 |

* See influence line curves (Figure A.19).

| | $U$ in m | $\dfrac{U}{2a}$ | $\operatorname{Sin}\dfrac{\pi u}{2a}$ | $\operatorname{Sin}\dfrac{2\pi u}{2a}$ | $\operatorname{Sin}\dfrac{3\pi u}{2a}$ | $\operatorname{Sin}\dfrac{4\pi u}{2a}$ |
|---|---|---|---|---|---|---|
| $U_1$ | 10.575 | 0.4273 | 0.9740 | 0.4411 | −0.7743 | −0.7917 |
| $U_2$ | 12.375 | 0.5000 | 1.0000 | 0.0000 | −1.0000 | 0.0000 |
| $U_3$ | 18.375 | 0.7424 | 0.7238 | −0.9989 | 0.6547 | 0.0954 |
| $U_4$ | 20.175 | 0.8152 | 0.5485 | −0.9172 | 0.9854 | −0.7307 |
| | | $\Sigma =$ | 3.2463 | −1.4751 | −0.1342 | 0.1617 |

$$M_{y_0} = P \times \frac{5.45}{12.375} [0.322 \times 3.2463 \times 1 + 0.0077 \times 0 \times 1.4751 + 0.027$$
$$\times 0.1342 \times 1 + 0]$$
$$= \frac{250}{4} \times 0.4404 \, (1.0453 + 0.0036)$$
$$= 28.9 \, \text{kNm/m}$$
$$M_{y_1} = 62.5 \times 0.4404 \, [0.22 \times 3.2463 \times 1.0 + 0.046 \times 0.1342 \times 1]$$
$$= 19.8 \, \text{kNm/m}$$
$$M_{y\alpha} = M_{y_0} + [M_{y_1} - M_{y_0}]\sqrt{\alpha} = 28.9 - 0.416 \times 9.1 = \underline{25 \, \text{kNm/m}}$$
$$\text{(sagging)}$$

*Local moments*

Calculated using Pucher chart.

Hogging = 26.2 kNm/m
Sagging = 14.4 kNm/m

*Local moment from DL + finish + edge beam*

$$W = 0.225 \times 18.5 + \left(\frac{1.73 + 23}{9.9}\right) = 8.2 \, \text{kN/m}$$

$$P_{\text{edge beam}} = 2.59 \, \text{kN}$$

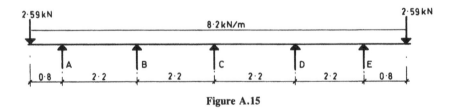

**Figure A.15**

*Moments* (Obtained by moment distribution method)

$$M_{\text{A[cant.mt.]}} = M_{\text{E[cant.mt.]}} = 4.7 \, \text{kNm/m (hogging)}$$
$$M_{\text{A-B}} = M_{\text{E-D}} = 4.7 \, \text{kNm/m (hogging)}$$
$$M_{\text{B-A}} = M_{\text{D-E}} = 2.95 \, \text{kNm/m (hogging)}$$
$$M_{\text{B-C}} = M_{\text{D-C}} = 295 \, \text{kNm/m (hogging)}$$
$$M_{\text{C-B}} = M_{\text{C-D}} = 3.47 \, \text{kN/m (hogging)}$$

Maximum sagging moment $= \dfrac{(2.95 + 3.47)}{2} - \dfrac{8.2 \times 2.2^2}{8} = \underline{1.75\,\text{kNm/m}}$

Maximum hogging moment $= \qquad\qquad\qquad\qquad = \underline{4.7\,\text{kNm/m}}$

Ult. moment: sagging $\quad = (1.15 \times 1.75 + 1.1 \times 14.4 + 1.1 \times 25)$
$\qquad\qquad\qquad\qquad\qquad \times 1.15 = \underline{52\,\text{kNm}}$

$\qquad\qquad$ hogging $\quad = (1.15 \times 4.7 + 1.1 \times 26.2) \times 1.15$
$\qquad\qquad\qquad\qquad\qquad\qquad = \underline{39.4\,\text{kNm}}$

Bottom reinforcement: T12 @ 125 c/c or equivalent

Top reinforcement: T10 @ 125 c/c or equivalent

Secondary reinforcement: T8 @ 250 c/c

*Check punching shear*

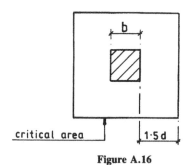

**Figure A.16**

Loaded area $= 1.1 \times \dfrac{250}{4} \times \dfrac{10^3}{1.1} = 62\,500\,\text{mm}^2$

$b = 250\,\text{mm}$

$d = 200 - 30 - \dfrac{12}{2} = 164\,\text{mm}$

$p = 4 \times 250 + 12 \times 164 = 2968\,\text{mm}$

Punching shear stress, $V = \dfrac{1.3 \times 62.5 \times 1000}{2968 \times 164} = 0.167\,\text{N/mm}^2$

Reinforcement provided at bottom of slab: T12 @ 125 c/c,
$\qquad\qquad\qquad\qquad\qquad\qquad$ As $= 905\,\text{mm}^2/\text{m}$

$\dfrac{100 \times 905}{1000 \times 164} = 0.55, \; V_c = 0.45\,\text{N/mm}^2 > V$

$\therefore$ OK (see reference [2], table 17).

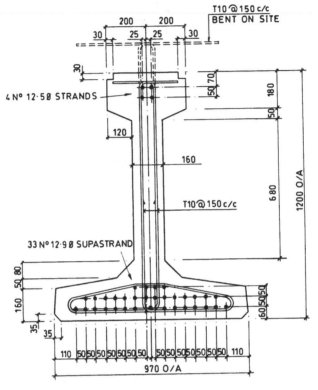

**Figure A.17**

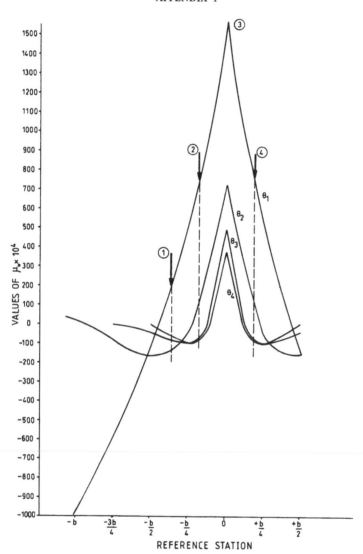

**Figure A.18**

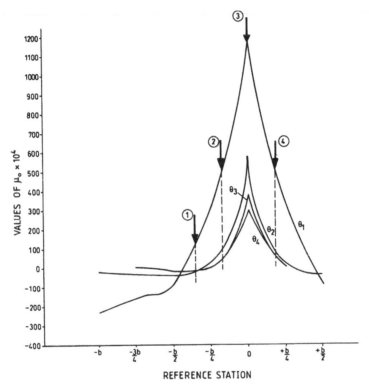

**Figure A.19**

*References and bibliography*

1. BS 5400, Part 4 (1990), *Steel Concrete and Composite Bridges. Part 4: Code of Practice for Design of Concrete Bridges*, British Standards Institution, London.
2. Rowe, R. E. *Concrete Bridge Design*, C.R. Books Ltd, 1962.
3. Rowe, R. E. *Concrete Bridge Design—Supplement Design Curves*, C.R. Books Ltd, 1962.
4. *Guide to the Structural use of Lightweight Aggregate Concrete*, The Institution of Structural Engineers and the Concrete Society, October 1987.
5. Abeles, P. W. and Bardhan-Roy, B. K., *Prestressed Concrete Designer's Handbook*, *3rd etn*, Viewpoint.
6. Higgins, G. E., Loe, J. A. and Howells, H., *The Lightweight Aggregate Concrete Bridge at Redesdale*, Transport and Road Research Laboratory Supplementary Report No. 788, 1983.
7. *Design and Cost Studies of Lightweight Concrete Highway Bridges*, Special Publication, Concrete Society, March 1986.

# Appendix 2

**Design of a cantilever roof beam of a grandstand structure using prestressed lightweight concrete**

This example is the design of the cantilever roof beams of the West Stand at St James Park, Newcastle, Design guidance in BS 8110, Part 2, is very inadequate for structural lightweight aggregate, especially for application to prestressed concrete design.

In this example, section 4 of BS 8110, Part 1, has been generally followed, with appropriate modification for lightweight concrete as suggested in the Institution of Structural Engineers Guide to *The Structural Use of Lightweight Aggregate Concrete*.

The example also includes the design of the end zone. Since there is no guidance in the British Code of Practice for end zone design of pretensioned members, the principles contained in CEB's *Bulletin d'Information*, No. 181 have been followed.

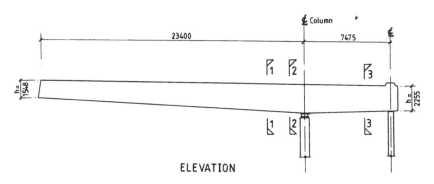

**Figure A.20**

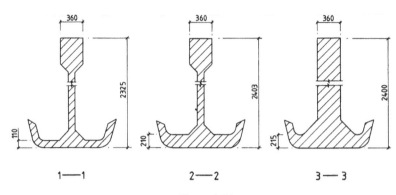

Figure A.21

*Basic load cases* (as shown in Figure A.22)

1. *Load Case 1*: 'Dead load'

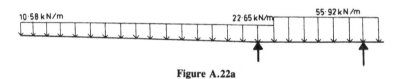

Figure A.22a

2. *Load Case 2*: 'Live load on cantilever only'

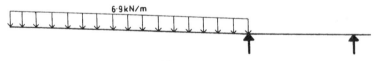

Figure A.22b

3. *Load Case 3*: 'Live load on the rear part only' (between support)

Figure A.22c

4. *Load Case 4*: 'Wind up front and down rear side'

Figure A.22d

5. *Load Case 5*: 'Wind down front and up rear point and back side'

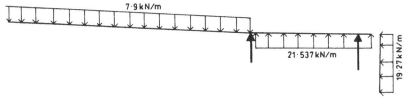

**Figure A.22e**

*Concrete*

At release, $f_{cu} = 50\,N/mm^2$, $E_c = 18\,kN/m^2$ ⎱ obtained from
At 28 days, $f_{cu} = 70\,N/mm^2$, $E_c = 19.125\,kN/m^2$ ⎰ manufacturer.
Density $= 1950\,kg/m^3$

*Prestressing wires*: $E = 195.1\,kN/mm^2$

*Dyform strands* (low relaxation):

15.2 mm, $F_u = 300\,kN$, 70% $F_u = 210\,kN$, Total 26 No. Area $= 165\,mm^2$

*Bridon strands* (low relaxation):

12.5 mm, $F_u = 165\,kN$, 70% $F_u = 115.5\,kN$, Total 8 No. Area $= 94.2\,mm^2$
9.3 mm, $F_u = 93.5\,kN$, 70% $F_u = 65.45\,kN$, Total 16 No. Area $= 52.3\,mm^2$
∴ Total prestressing force $= 26 \times 210 + 8 \times 115.5 + 16 \times 65.45$
$= 7431.20\,kN$

CG of prestressing force $= 1658.425 - 923.145 = 735.28\,mm$

The following calculation relates to Section 2-2 only.

*Section at 515 mm from ₵ of column* (Section 2-2)

*Properties*:

| | Concrete section | Entire section (taking bonded steel into consideration) |
|---|---|---|
| $A =$ | $675\,430\,mm^2$ | $729\,535.561\,mm^2$ |
| $\bar{y}_B =$ | $923.145\,mm$ | $976.941\,mm$ |
| $\bar{y}_T =$ | $1479.855\,mm$ | $1426.059\,mm$ |
| $I_{XX} =$ | $508\,520.601 \times 10^6\,mm^4$ | $575\,087.763 \times 10^6\,mm^4$ |
| $Z_B =$ | $550.857 \times 10^6\,mm^3$ | $588.662 \times 10^6\,mm^3$ |
| $Z_T =$ | $343.629 \times 10^6\,mm^3$ | $403.271 \times 10^6\,mm^3$ |

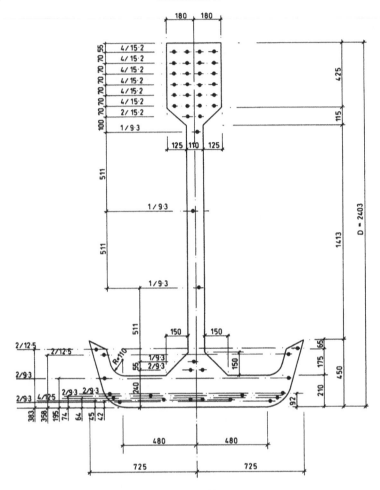

**Figure A.23**

*Limit state of collapse condition*

*Hogging bending moment*

$M_{u[1.2(DL + LL + WL)]} = \underline{-10127.362\,kNm}$

Strands at the top: total $F_u = 0.87 \times 300 \times 26$ No. $= \underline{6786\,kN}$

CG $= 248.846\,mm$ from the top of beam

$\therefore d = 2403 - 248.846 = 2154.154\,mm$

$d_n = \dfrac{6786 \times 10^3}{0.45 \times 70 \times 1150} = 187.33\,mm$

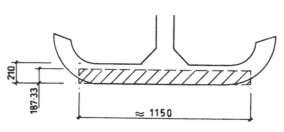

**Figure A.24**

$$\therefore l_a = (2154.154 - \frac{187.33}{2}) = 2060.5\,\text{mm}$$

$$\frac{f_{pu}A_{ps}}{f_{cu}bd} = \frac{6786 \times 10^3}{70 \times 1150 \times 2154.154} = 0.039$$

$$\frac{f_{pe}}{f_{pu}} = 0.7 \times 0.587 = 0.41 \longrightarrow \frac{f_{pb}}{0.87f_{pu}} = 1.0$$

$$\therefore M_R = 6786 \times 2060.5 \times 10^{-3} = \underline{13\,982.6\,\text{kNm}} > M_u \quad \therefore \text{OK}$$

*Sagging bending moment*

$M_{u(1.0DL + 1.4W\uparrow)} = 3584.25\,\text{kNm}$

Strands at the bottom: total $F_u = 0.87(13 \times 93.5 + 8 \times 165) = 2206\,\text{kN}$

CG = 196.746 mm from the bottom of beam

$$\therefore d = 2403 - 196.746 = 2206.254\,\text{mm}$$

$$d_n = \frac{2206.254}{0.45 \times 70 \times 360} = 194.53\,\text{mm}$$

$$\therefore l_a = 2206.254 - \frac{194.53}{2} = 2109\,\text{mm}$$

$$\frac{f_{pu}A_{ps}}{f_{cu}bd} = \frac{2206 \times 10^3}{70 \times 360 \times 2206.254} = 0.04,$$

$$\frac{f_{pe}}{f_{pu}} = 0.7 \times 0.587 = 0.41 \longrightarrow \frac{f_{pb}}{0.87f_{pu}} = 1.0$$

$$\therefore M_R = 2206 \times 2109 \times 10^{-3} = \underline{4652\,\text{kNm}} > M_u \quad \therefore \text{OK}$$

*Check for service condition*

*Prestressing tendons*

A = 5880.4 mm$^2$

$\bar{y}_B$ = 1658.425 mm

$\bar{y}_T$ = 744.575 mm

$e$ = 735.28 mm

Initial prestressing force = 7431.2 kN.

*Stresses*

$$f_t = -\frac{7431200}{675430} - \frac{7.4312 \times 735.28}{343.629} = -26.9\,\text{N/mm}^2\,(\text{C})$$
(after instantaneous loss, this will $< f_{cu}/2$ (at release))

$$f_b = -\frac{7431200}{675430} + \frac{7.4312 \times 735.28}{550.857} = -1.083\,\text{N/mm}^2\,(\text{C})$$

Stress at CG of steel, $f_{co} = -1.083 + \left[\frac{(-26.9) - (-1.083)}{2403}\right] \times 1658.425$

$$= -18.903\,\text{N/mm}^2\,(\text{C})$$

*Losses* 41.3%
For calculation of losses, the following have been considered:

Relaxation = 1.1% of initial force
Shrinkage = $300 \times 10^{-6}$
Creep = $72 \times 10^{-6}$ per N/mm$^2$

After losses (purely due to prestress):
$f_t = -15.8\,\text{N/mm}^2\,(\text{C})$
$f_b = -0.64\,\text{N/mm}^2\,(\text{C})$

1. *Maximum hogging moment*

$M_{[\text{DL}+0.8(\text{LL}+\text{WL})]} = -7640.45\,\text{kN m}$

*Resultant stresses* (prestress and load):

$$f_t = -15.80 + \frac{7640.45}{403.271} = 3.15\,\text{N/mm}^2\,(\text{T}) < 4.11\,\text{N/mm}^2\,(\text{OK})$$

$$f_b = -0.64 - \frac{7640.45}{588.662} = -13.62\,\text{N/mm}^2\,(\text{C}) < 23.1\,\text{N/mm}^2\,(\text{OK})$$

2. *Maximum sagging moment*

$M_{(1.0\text{DL}+1.0\text{W}\uparrow)} = 1264.44\,\text{kN m}$

*Resultant stress* (prestress and load):

$$f_t = -15.8 + \frac{1264.44}{403.271} = -12.66\,\text{N/mm}^2\,(\text{C}) < 23.1\,\text{N/mm}^2\,(\text{OK})$$

$$f_b = -0.64 - \frac{1264.44}{588.662} = -2.80\,\text{N/mm}^2\,(\text{C}) < 23.1\,\text{N/mm}^2\,(\text{OK})$$

Maximum tensile stress allowable = $0.7 \times 7.3 \times 0.8 = \underline{4.11\,\text{N/mm}^2}$

Maximum compression stress allowable $= 0.33 \times 70 = \underline{23.1\,\text{N/mm}^2}$

*Shear design* (Cl 4.3.8)

(A) Load case: $1.2(1 + 2 + 5)$: $S \cong 811\,\text{kN}$, $M \cong -10\,127.36\,\text{kN m}$
(B) Load case: $(1.0 \times 1 + 1.4 \times 4)$: $S \cong -367\,\text{kN}$, $M \cong 3584\,\text{kN m}$

$M_0$ at top $= 0.80 \times 15.80 \times 403.271 = 5543\,\text{kN m}$
$M_0$ at bottom $= 0.80 \times 0.64 \times 588.662 = 328\,\text{kN m}$
$M = -10\,127.36\,\text{kN m} > M_0$ (5543 kN m)   $\therefore$ Section cracked
$M = 3584\,\text{kN m} > M_0$ (328 kN m)   $\therefore$ Section cracked

*(A)*

$$f_{cp} = -0.64 - 13.615 \times \frac{923.145}{2403} = -5.87\,\text{N/mm}^2$$

1. $V_{co} = 0.67 \times 110 \times 2403\sqrt{(2.008^2 + 0.8 \times 5.87 \times 2.008)}$
   $\times 10^{-3} = 650\,\text{kN}$

   $f_{pe} = 0.7 \times 0.587 f_{pu} = 0.411 f_{pu}$
   $d = 2403 - 248.846 = 2154.154\,\text{mm}$

   $\dfrac{100 \times 26 \times 165}{110 \times 2154.154} = 1.80$   $\therefore V_c = 0.9 \times (0.8) = 0.72\,\text{N/mm}^2$

2. $V_{cr} = (1 - 0.55 \times 0.411) \times 0.72 \times 110 \times 2154.154 \times 10^{-3}$
   $+ 5543 \times \dfrac{811}{10\,127.36}$

   $= 576\,\text{kN}$

   $V_c + 0.4 b_v d = 576 + 0.4 \times 110 \times 2154.154 \times 10^{-3} = 671\,\text{kN}$
   $V$ (811 kN) $> (V_c + 0.4 b_v d)$
   $S_v = \dfrac{0.87 \times 250 \times 157 \times 2154.154}{(811 - 671) \times 10^3} = 525\,\text{mm}$

*Provide R10 (2 legs) at 200 c/c*

*Shear check*

*(B)*

1. $V_{co} = \underline{650\,\text{kN}}$

   $d = 2403 - 84.34 = 2318.66\,\text{mm}$ $\left.\begin{array}{l}\\ \\\end{array}\right\}$ (Consider strand in the
   $A_s = 10 \times 52.3 + 4 \times 94.2 = 900\,\text{mm}^2$ bottom 210 flange only)

$$\frac{100 A_s}{b_v d} = \frac{100 \times 900}{210 \times 2318.66} = 0.185 \quad \therefore V_c = 0.421 \, \text{N/mm}^2 \times 0.8$$
$$= 0.337 \, \text{N/mm}^2$$

$f_{pe} = 0.411 f_{pu}$

$$\therefore V_{cr} = (1 - 0.55 \times 0.411) \times 0.337 \times 210 \times 2318.66 \times 10^{-3}$$
$$+ 328 \times \frac{367}{3584}$$
$$= \underline{160.6 \, \text{kN}}$$

$V_c + 0.4 b_v d = 160.6 + 0.4 \times 210 \times 2318.66 \times 10^{-3} = \underline{355 \, \text{kN}}$

$V \, (367 \, \text{kN}) > (V_c + 0.4 b_v d)$

$$S_v = \frac{0.87 \times 250 \times 157 \times 2318.66}{(367 - 355) \times 10^3} = 6598 \, \text{mm}$$

*Provide R10 (2 legs) at 200 c/c*

*End zone design (Bulletin d'Information, No. 181)*

*Transmission length (BS 8110, Cl 4.10)*

$l_t = \dfrac{K_t \phi}{\sqrt{f_{ci}}}$  Strands 15.2 mm, $K_t = 360$, $f_{ci} = 50 \, \text{N/mm}^2$
      Strands 12.5 mm, $K_t = 240$

$l_t \, (15.2 \, \text{mm}) = 774 \, \text{mm}$

$l_t \, (12.5 \, \text{mm}) = 424 \, \text{mm}$

*Bursting stresses*:

Symmetrical Prism Rule: $h_{bs} = l_{bs} = 2 \times 0.15 \times h$

Front of beam: $h = 1548 \, \text{mm} \quad \therefore h_{bs} = 464.4$

$$c_{bs} = \sqrt{1 + \left(\frac{0.6 l_t}{h_{bs}}\right)^2} = \sqrt{1 + \left(\frac{0.6 \times 774}{464.4}\right)^2} = 1.414$$

Length of prism $= c_{bs} \times l_{bs} = 1.414 \times 464.4 = \underline{656.76 \, \text{mm}}$

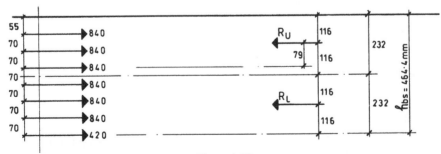

**Figure A.25**

CG from top $= 248.846$ mm

$R_L \qquad = 3126.446$ kN

$R_u \qquad = 2333.534$ kN

$-M_{Ru} \ = -840 \times (116 - 55) = -51.24 \times 10^6$ N mm

$-M_{RL} \ = -420 \times (116 + 11) - 840(127 - 70) = -101.22 \times 10^6$ N mm

$+M_{max} = 2333.534 \times 79 - 840 \times 140 - 840 \times 70 = +7.95 \times 10^6$ N mm

$Z \qquad = \frac{1}{2} \times 628.57 = 314.285$ mm

$P_{bs} \qquad = \dfrac{101.22 \times 10^6}{314.285} \times 10^{-3} = \underline{322.064 \text{ kN}}$

Maximum bursting stress $= \dfrac{P_{bs}}{\frac{2}{3} \times \frac{3}{4} \times l_s \times b}$

$\qquad\qquad\qquad\qquad = \dfrac{322064 \times 2}{656.76 \times 360} = 2.72 \text{ N/mm}^2$

Required links $= \dfrac{322064}{0.67 \times 0.87 \times 250} = 2210.08/\text{mm}^2$

*R10 links, 14 No. in 657 mm length*

*At the rear*:

$h \quad = 2255$ mm

$h_{bs} = 0.3 \times 2255 = 676.5$ mm

$c_{bs} = \sqrt{\left[1 + \left(\dfrac{0.6 \times 774}{676.5}\right)^2\right]} = 1.213$

Length of prism $= 1.213 \times 676.5 \cong 821$ mm

Internal lever arm $= \frac{1}{2} \times 821 = 410$ mm

$P_{bs} = \dfrac{M}{Z} = \dfrac{101.22 \times 10^6}{410} \times 10^{-3} = 246.71 \text{ kN}$

Maximum bursting stress $= \dfrac{246\,710 \times 2}{821 \times 360} = 1.67 \text{ N/mm}^2$

Required links $= \dfrac{246\,710}{0.67 \times 0.87 \times 250} = 1693.0 \text{ mm}^2$

*R10 links, 11 No. in 821 mm length*

*Spalling forces and stresses*

*Front section $D = 1548$ mm, $b_v = 360$ mm*

*Properties* (concrete):

$\bar{y}_B = 645.334$ mm, $\bar{y}_T = 902.666$ mm, $I_{XX} = 157\,545.82 \times 10^6$ mm$^4$

$Z_B = 244.131 \times 10^6 \, \text{mm}^3$, $Z_T = 174.534 \times 10^6 \, \text{mm}^3$.

*CG Wires:*

$\bar{y}_B = 1015.16 \, \text{mm}$, $\bar{y}_T = 532.838 \, \text{mm}$, $e = 369.826 \, \text{mm}$

$$f_t = \frac{-7431200}{703155} - \frac{7.4312 \times 369.826}{174.534} = -26.314 \, \text{N/mm}^2$$

$$f_b = \frac{-7431200}{703155} + \frac{7.4312 \times 369.826}{244.131} = +0.689 \, \text{N/mm}^2$$

At CG steel, $f_s = +0.689 - 17.708 = -17.019 \, \text{N/mm}^2$

After losses, residual stress % = 0.605

$f_t = 0.605 \times (-26.314) = -15.929 \, \text{N/mm}^2$

$f_b = 0.605 \times (+0.689) = +0.417 \, \text{N/mm}^2$

Depth of tensile zone $= 1548 \times \dfrac{0.417}{16.346} = 39.491 \, \text{mm}$

Equilibrium of bottom prism, when SF = 0 at depth of $2 \times 39.491$ = 78.982 mm

Length of prism in a pretensioned beam increased by

$$c_{sp} = \sqrt{\left[1 + \left(\frac{0.6 \times 424}{1548}\right)^2\right]} = 1.013 \quad (l_{t(12.5 \, \text{mm})} = 424 \, \text{mm})$$

$L = 1.013 \times 1548 = 1568.76 \, \text{mm}$

$Z = 0.5 \times 1568.76 = 784.38 \, \text{mm}$

$$F_{sp} = \frac{M}{Z} = \frac{0.5 \times 0.417 \times 39.491 \times \text{⁴/₃} \times 39.491 \times 950}{784.38} \qquad b \cong 950 \, \text{mm}$$

$$= 525 \, \text{N}$$

Maximum tensile concrete stress at the end of beam:

$$\sigma_{sp} = \frac{F_{sp}}{\frac{1}{2}b \times \frac{1}{4}h} = \frac{525.0 \times 8}{950 \times 1565.318}$$

$$= 0.003 \, \text{N/mm}^2$$

$$\approx 0$$

# Appendix 3

**Design of lightweight concrete prestressed double-T unit construction for 4 hours' fire resistance**

This example illustrates the application of analytical techniques for the assessment of fire resistance of flexural member and, in fact, refers to the actual construction at Clifton Down Shopping Centre at Bristol. The long span (15.6 m) suspended ground floor (over the basement) needed 4 hours' fire resistance. For normal temperature condition the floor required 700 deep double-T units in lightweight aggregate concrete, with structural topping of minimum 50 mm thickness.

The floor was designed for normal temperature condition first and then analytically checked for 4 hours' fire resistance in accordance with the principle and procedure laid down in the *Design and Detailing of Concrete Structures for Fire Resistance* – an Interim Report by the joint Committee of the Institution of Structural Engineers and the Concrete Society. As the actual variation in the loading, span, detailing and the bearing conditions can be duly taken into account, the analytical method provides a more reliable prediction of fire resistance of a member.

*Clifton Down Shopping Centre Complex, Bristol*

The suspended ground floor constructed at above consists of 700 deep prestressed double-T units with 50 mm minimum structural topping to span 15.6 m carrying shop floor load as follows:

1. Own weight precast = 5.95 kN/m⌉ DL = 5.95 + 4.32 + 3.6
2. *In-situ* topping = 4.32 kN/m│   = 13.87 kN/m
3. Finishes = 3.60 kN/m⌋
4. Live load: (i)  10.5 kN/m² = 25.20 kN/m
   (ii)   5.0 kN/m = 12.00 kN/m

The prestressed units are in lightweight concrete using Lytag aggregate.

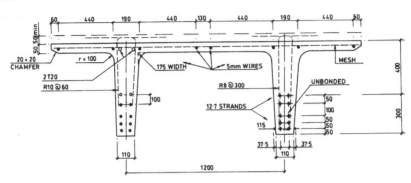

**Figure A.26** Cross-section of double-T.

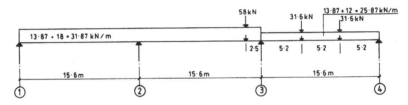

**Figure A.27** Service load diagram (normal temperature).

*Design for normal temperature condition*

*(A)   Serviceability condition:*

Residual stresses due to full service load (DL + LL) and prestressing:

$f_{tp}$ (stress at top fibre)      = −8.10 N/mm² (C)
$f_{bp}$ (stress at bottom fibre) = +4.94 N/mm² (T)

(OK for Class 3 design, corresponding to 0.2 mm crack width)

*(B)   Ultimate condition*

$M_{(ultimate)}$   =   976.64 kN m
$M_{(resistance)}$ = 2266.69 kN m

$$FS \qquad = \frac{2266.69}{976.64} = 232 \quad \therefore OK$$

*Calculations for fire resistance*

The calculations follow the procedure shown in reference [2] (p. 193).

CG of strands = 170 mm above bottom of DT ribs

Average side cover to centre of steel $= \dfrac{127 - 40}{2} = 43.5\,\text{mm}$

Bottom cover (to centre of steel) $= 49.95\,\text{mm}$

Partial safety factor for load: DL $= 1.05$

$\qquad\qquad\qquad\qquad\qquad$ LL $= 1.00$

Partial safety factor for material: Concrete $= 1.30$

$\qquad\qquad\qquad\qquad\qquad\qquad$ Steel $\quad = 1.00$

*Double-T, D700 Design for 4-hours' fire resistance*
*Loading*

DL: UDL $\quad = 1.05 \times (5.95 + 4.32 + 3.6) = 14.56\,\text{kN/m}$

$\quad$ Point load $= 1.05 \times 58 \qquad\qquad = 60.90\,\text{kN}$

$\quad$ Point load $= 1.05 \times 31.6 \qquad\quad = 33.18\,\text{kN}$

LL: UDL $\quad = 7.5 \times 2.4 \qquad\qquad = 18.00\,\text{kN/m}$

$\quad$ UDL $\quad\;\; = 5.0 \times 2.4 \qquad\qquad = 12.00\,\text{kN/m}$

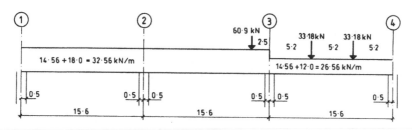

**Figure A.28**

*Free BM at mid-span*

$$M_{1-2} = 32.56 \times 15.6^2 \times \frac{1}{8} = 990.48\,\text{kN m}$$

$$M_{2-3} = 990.48 + 60.9 \times 2.5 \times 7.8 \times \frac{1}{15.6} = 1066.61\,\text{kN m}$$

$$M_{3-4} = 26.56 \times 15.6^2 \times \frac{1}{8} + 33.18 \times 5.2 = 980.49\,\text{kN m}$$

*Moment of resistance of section at mid-span* (after 4 hours' fire exposure)

CG of strands $= 170\,\text{mm}$ from bottom of rib

Average side cover to centre of steel $= 43.5\,\text{mm}$

(reference [2], Table 15e)

Temp. at 43.5 mm inside vertical face of rib (after 4 hours' exposure) = 780°C (for normal concrete).

For lightweight concrete, temperature = 0.8 × 780
$$= 624°C$$

(reference [2], Fig. 13)

$$\frac{\text{Strength of steel at } 624°C}{\text{Strength of steel at } 20°C} = 0.155$$

∴ Moment of resistance of section at mid-span

20 No. 12.7 dia. strands, ch. strength (normal temp.) = 20 × 209
$$= 4180 \text{ kN}$$

∴ $F_t = 0.155 × 4180 = 647.9 \text{ kN}$

(From Figs 14a and 14b, reference [2]) top slab being 100 mm thick, temp. of RC slab at 100 mm away from exposed face = <380°C. Hence (from Fig. 12, [reference [2]), 100% strength retained in concrete at top surface.

$$\text{Uniform compression } c = \frac{0.67 × 30 × \overset{\text{(Age factor)}}{1.2}}{1.3} = 18.55 \text{ N/mm}^2$$

$$\text{Depth of concrete in compression} = \frac{647.9 × 10^3}{2400 × 18.55} = 14.6 \text{ mm}$$

$$M_u = 647.9 × 10^3 \, (750 - 170 - 9.3) × 10^{-6} = 369.7 \text{ kN m}$$

Average negative moments over supports required = 1066.61 × 369.7
$$= 696.91 \text{ kN m}$$

Consider  T16 at 220c/c on grids 1 and 4
and        T16 at 150c/c on grids 2 and 3

$M_u$ at the face of support,  i.e. $x = 0.5$ m

Grids 1 and 4:

Effective $d = 800 - 30 = 770$ mm

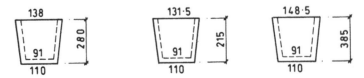

**Figure A.29**

Maximum depth in compression $= \frac{1}{2}d = 385\,\text{mm}$

Neglecting 10 mm thickness of exposed face find mean temperature in concrete (from Fig. 15f, reference [2])

At 10 mm,   $T = 1040 \times 0.8 = 832°\text{C}$
At 55 mm,   $T = \phantom{0}740 \times 0.8 = 592°\text{C}$
At 74.5 mm, $T = \phantom{0}710 \times 0.8 = 568°\text{C}$

Mean $T = \frac{1}{3}(832 + 592 + 568) = 664°\text{C}$

From Fig. 12 (Ref. 2), percentage strength retained: $= 0.4 + \dfrac{136}{300} \times 0.6$

$$= 0.672$$

$\therefore$ Uniform $c = \dfrac{0.67 \times 60 \times 1.2}{1.3} \times 0.672 = 25\,\text{N/mm}^2$

$\therefore$ Max. $F_c = 2 \times \frac{1}{2}(91 + 128.5) \times 375 \times 25 \times 10^{-3} = 2058\,\text{kN}$

Max. $F_t = 11\,T_{16} = 11 \times 201 \times 460 \times 10^{-3} \phantom{00} = 1017\,\text{kN}$

Equivalent $A_c = \dfrac{1017 \times 10^3}{25} = 40\,682\,\text{mm}^2$

Consider $A_c \phantom{0} = \phantom{000000000}$ $A_c \phantom{0} = 2 \times \frac{1}{2}(91 + 111.5) \times 205$
$$= 41\,513\,\text{mm}^2$$

$$\text{CG} = 10 + \frac{205}{3} \times \frac{314}{202.5} = 116\,\text{mm}$$

$\therefore M_u = 1017 \times (770 - 116) \times 10^{-3} = \underline{\underline{665\,\text{kN m}}}$

*Grids 2 and 3*

$F_t = \text{T16 at 150 c/c} = 15\,T\,16$
$$= 15 \times 201 \times 460 \times 10^{-3} = 1386.9\,\text{kN}$$

Equivalent $A_c = \dfrac{1386.9 \times 10^3}{25} = 55\,476\,\text{mm}^2$

Consider 280 mm depth of concrete in compression $A_c = \phantom{0000}$ $A_c \phantom{0} = 2 \times \frac{1}{2}(91 + 118) \times 270$
$$= 56\,430\,\text{mm}^2$$

$$\text{CG} = 10 + \frac{270}{3} \times \frac{327}{209} = 150.8$$

$\therefore M_u \approx 1386.9 \times (770 - 150.8) \times 10^{-3} = \underline{\underline{858.8\,\text{kN m}}}$

*Span 1–2*

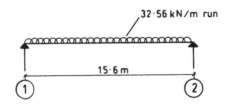

**Figure A.32**

$M_u$ (span)          $= 369.7 \, \text{kN m}$

$M_u$ (Support 1) $= 665 \, \text{kN m}$ ⎱ Av. $= 761.9 \, \text{kN m}$
   (Support 2) $= 858.8 \, \text{kN m}$ ⎰ $\Sigma = 1131.6 \, \text{kN m}$

Free BM $= 990.48 \, \text{kN m} < M_u \, (1131.6 \, \text{kN m})$  $\therefore$ OK

*Probable moment diagram (after 4 h exposure)*

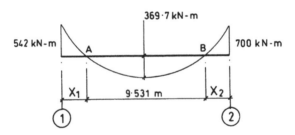

**Figure A.33**

Average support moment $= 990.48 - 369.7 = 620.78 \, \text{kN m}$

$$M_1 = \frac{620.78}{761.9} \times 665 = 542 \, \text{kN m}$$

$M_2 = 0.81 \times 858.8 = 700 \, \text{kN m}$

Let $x$ be the distance between points of contraflexure,
$w =$ Design fire condition load $= 32.56 \, \text{kN/m}$

$$\frac{wx^2}{8} = 369.7 \qquad \therefore x = 9.531 \, \text{m}$$

$$R_A = R_B = 32.56 \times \frac{9.531}{2} = 155.2 \, \text{kN}$$

*To find $x_1$*

$542 = 155.2 \times x_1 + 32.56x^2/2$

$\therefore x_1 = 2.718\,\text{m}$

$x_2 = 15.6 - 2.718 - 9.531$

$= 3.351\,\text{m}$

*Check shear*

*Support 1*

$$S_1 = 32.56 \times \frac{15.6}{2} - \frac{700 - 542}{15.6} = 254 - 10.1 = 243.8\,\text{kN}$$

Shear at face of support (500 mm from Grid 1)

$= 243.8 - 32.56 \times 0.5$

$= \underline{227.52\,\text{kN}}$

Moment at face of support

$= 155.2(2.718 - 0.5) + 32.56 \times \dfrac{2.218^2}{2}$

$= \underline{424.4\,\text{kN m}}$

$d = 770\,\text{mm}$

$$k_v = \frac{M}{Vd} = \frac{424.4}{227.52 \times 0.77} = 2.42 > 2$$

$\therefore$ Section 8.2.2.2(iv) applies (reference [2])

Shear span, $l_{vx} = \dfrac{M}{V} = \dfrac{424.4}{227.52} = 1.865.$

$$M_{cx} = 0.875\,dl_{vx}\left[0.342b_1 + 0.3\frac{M_{ux}}{d^2}\sqrt{\frac{2}{l_{vx}}}\right]$$

$$\times \sqrt[4]{\frac{16.66}{\rho f_y \psi_s T^\circ}}$$

$Z = 770 - 116 = 654\,\text{mm}$

$b_1 = 2\left(\dfrac{91 + 111.5}{2}\right) = 202.5\,\text{mm}$ (2 ribs)

$M_{ux} = 665\,\text{kN m}$

$$\rho = \frac{\text{Area of tensile reinforcement}}{\text{Area of concrete to effective depth}} = \frac{11 \times 201}{330497 + 2400 \times 20}$$

$= 0.00584$

$$M_{cx} = 0.875 \times 770 \times 1865 \left[ 0.342 \times 202.5 + 0.3 \times \frac{665 \times 10^6}{770^2} \sqrt{\frac{654}{1865}} \right]$$

$$\times \sqrt[4]{\frac{16.66}{0.00584 \times 460}}$$

$$= (1\,256\,543.8[69.26 + 199.26] \times 1.578) \times 10^{-6}$$

$$= 532\,\text{kN m} > 424.4\,\text{kN m}$$

Hence OK

*Hanger bars required at point of contraflexure* (reference [2], Table 15e)

Shear $= 155.2\,\text{kN}$

Cover to links $= 25\,\text{mm}$      Temp. of steel $= 900 \times 0.8$
$$= 720°\text{C}$$

$$\frac{\text{Strength of steel at T°C}}{\text{Strength of steel at 20°C}} = 0.05 + 0.95 \times \frac{80}{500} = 0.202$$

$$T_{10} \text{ links required} = \frac{155.2 \times 10^3}{(2 \times 78) \times 460 \times 0.202} = \begin{array}{l} 10.71\,\text{Nos/2 Ribs} \\ \text{i.e. by 6 Nos/Rib} \end{array}$$

# Index

abrasion resistance  35
admixtures  82, 86–92
aggregate properties
  bulk density  21, 78–79
  loss on ignition  22
  particle density  1, 3
  particle shape  2, 21
  water absorption  21–23, 78–80, 85
aggregates
  Aglite  4, 17, 112, 123
  Basalite  130
  Brag  4, 17
  Fibo  see Leca
  Foamed slag  7–8, 21
  Granulex  7, 12–14
  Haydite  4, 170, 172
  Leca  7–10, 36
  Liapor  see main entry
  Lionite  124, 132
  Litecrete  121, 124
  Lytag  see main entry
  Materialite  117, 124
  Pellite  7, 12–13
  Pumice  106
  Russlite  4, 17
  Shalite  132
  Solite  4, 12
Aglite  4, 17, 112, 123
American standards
  Arctic structures  48
  buildings  20, 48–51, 54–58, 61–70
  offshore platforms  48
Arctic structures  48
Atlantus  172
Australia Square  120–121
Australian standards  19–20, 49–51, 54, 58, 61
Avelengo bridge  164–165

Basalite  130
Blackfriars bridge  166
blastfurnace slag  4, 21
blockwork  136–141
BMW administrative building  122, 184–185
Boknasudet bridge  182–183
Brag  4, 17

bridges
  Avelengo, Italy  164–165
  Blackfriars, London  166
  Boknasudet  182–183
  design example  195–218
  Ealing Broadway Centre  150
  Friarton  165
  Herne Bridge, Oundle  142
  Koningspleijbrug, Netherlands  151, 155, 157–161, 182
  Redesdale  162
  Ulft, Netherlands  162–164
British Library  125
British standards
  aggregates  21
  bridges  47, 50–51, 54–66, 70–71
  buildings  46, 50–51, 53–70
  marine and coastal structures  47
buildings
  Australia Square  120–121
  BMW administrative building  122, 184–185
  British Library  125
  Canary Wharf  88–91, 111, 126
  Central Square  124
  Chinese Embassy, London  120–121
  Clifton Down Shopping Centre  184–187, 229–237
  Commercial Centre Tower  124
  Guy's Hospital  117–119
  Kensington Maintenance Depot  114
  Lake Point Tower  123–124
  Lloyds Building, London  132–133
  Marina City  116–117
  National Theatre, Tokyo  132
  NLA Tower, Croydon  118–119
  One Shell Plaza  17, 120, 183–184
  Pantheon  106
  Pimlico Secondary School  124
  Raymond Hillard Centre  117
  Roxburgh County Offices  115
  Royal Exchange  142
  Scotstoun House, South Queensferry  131
  Sheraton Park Tower  124
  Standard Bank, Johannesburg  122–123

buildings *contd*
   Student Union, San José   130
   Torre Picasso   125

Calgary grandstand   189–191
Canary Wharf   88–91, 111, 126
carbonation   36–37
Central Square   124
Chinese Embassy, London   120–121
clay   4, 8–10
Clifton Down Shopping Centre   184–187,
   229–237
columns   63–64, 72
Commercial Centre Tower   124
compaction   93, 101
Concrete Society studies
   bridges   101, 150–152
   buildings   101–102, 143–145
construction
   compaction   93, 101
   curing   95–96
   finishing and finishes   93–95, 112, 124,
     130–131
   fixing   95, 100, 154
   floatation of particles   93–94
   placing   93, 110–111
   pour sizes   93
   power floating   94–95, 97, 112
   profiled metal decking   93–94, 97,
     126–129, 142
   pumping   86–92
   repair   100
   slip forming   111–112, 117
   steam curing   170
   thermal cracking   69–70
   transport   85–86
   vacuum de-watering   97
   vacuum soaking   79–80, 90–92
   water control and addition   78–80,
     85–86, 92, 104–105, 111, 165
   yield   85–96
corrosion of steel   37
cover to reinforcement   51–54, 66–67
cryogenic properties   39
curing   95, 96

deflections   61–63, 72
density
   aggregates   1, 3
   air dry   23, 98–99
   classification classes   20, 50
   design values   50
   fresh concrete   23–24, 98–99
   oven dry   20, 23, 50, 58–59
   saturated   23
design requirements   *see* British
   standards, etc.
durability

abrasion resistance   35
behaviour in service   40–41, 51–52,
   107–108, 162, 173
carbonation   36–37
chemical resistance   35
chloride penetration   36
corrosion of steel   37, 173
cover to reinforcement   51–52
freeze–thaw behaviour   34–35
permeability   35–36
resistivity   36
water absorption   35

economics   101–102, 113, 120, 127, 135,
   143–148, 150–153, 159, 175–176,
   182–183, 193
European standards   48–49, 55–58

Fibo   *see* Leca
finishing and finishes   93–95, 112, 124,
   130–131
fire
   design example   229–237
   design requirements   53–54, 108–109,
     116, 185
   experimental evidence   39–40, 128
   in-service behaviour   54, 99, 127
fixing   85, 100, 154
flexure   55, 72
floatation of particles   93–94
foamed slag   7–8, 21
freeze–thaw resistance   95–96
Friarton bridge   165

Genoa floating dock   175–176
grandstands
   Calgary   189–191
   Newcastle   191–193, 219–228
   Twickenham   187–189
Granulex   7, 12–14
Guy's Hospital   117–119

Haydite   4, 170, 172

Japanese standards   20, 49–51, 54–58,
   61–70

Kensington Maintenance Depot   114
Koningspleijbrug   151, 155, 157–161, 182

Lake Point Tower   123–124
Leca
   concrete properties   10, 36
   manufacturing process   8–10
   particle properties   7, 9
Liapor
   applications   122, 136, 152, 180–184
   concrete properties   5, 36

manufacturing process 12–14
particle properties 7, 15
Lionite 124, 132
Litecrete 121, 124
Lloyds Building, London 132–133
Lytag
  applications 90–91, 112, 114, 115,
    118–120, 124–126, 142, 162–165, 182,
    185, 187–189, 191–193
  concrete properties 4, 7, 36, 54, 109
  manufacturing process 10–12
  particle properties 11
  workability control 104–105

manufacture of aggregates 5–14
Marina City 116–117
Materialite 117, 124
mix design
  admixtures 82
  aggregate content 83
  cement content 23, 83, 125
  cement replacements 81–83
  control of moisture content 23, 80,
    83–85
  lightweight fines 81, 109, 125, 169–170
  pumping 81–82
mixing 84–85

National Theatre, Tokyo 132
Newcastle grandstand 191–193, 219–228
NLA Tower, Croydon 118–119
no-fines concrete 19
Norwegian standards 20, 48, 50–51,
  54–58, 60–70

offshore structures
  America 48
  comparison with NWC 30
  durability 40, 48
  proposed oil storage tanks 176–180
One Shell Plaza 17, 120, 183–184
oxygen permeability 36

Pantheon 106
Pellite
  concrete properties 7
  manufacturing process 12
  particle properties 7, 13
PFA 4, 5, 14–16 *see also* Lytag
Pimlico Secondary School 124
placing 93, 110–111
pour sizes 93
power floating 94–95, 97, 112
precast concrete
  cladding panels and wall units 115,
    129–132, 176–180
  floors and beams 113, 121–123, 125,
    130, 176–180, 185, 188–193

standard systems 132–136
prestressed concrete
  design example 219–228
  design requirements 66–69, 73
  structures 113–114, 130, 147, 161,
    164, 168–169, 176–183, 188–193
production processes
  foamed slag 7–8
  pelletised expanded blastfurnace
    slag 12
  rotary kiln 8–10, 12–14
  sinter strand 10–11
profiled metal decking 93–94, 97, 126–
  129, 142
properties of fresh concrete
  density 23–24, 98–99
  water absorption 21–23, 78–80
  workability 30, 83, 97–98
properties of hardened concrete
  abrasion resistance 35
  acoustic insulation 39
  bond and anchorage 33–34
  carbonation 36–37
  chemical resistance 35
  compressive strength 19–20, 24–26,
    45–51, 78, 97, 109–110, 112, 118–120,
    125, 136, 162, 165, 168–169
  creep 33, 69
  cryogenic 39, 173–175
  density 19–20, 23, 45–50, 109–110,
    169
  elastic modulus 32–33, 61–62
  fatigue 34
  fire resistance 39–40, 99
  freeze–thaw resistance 34–35, 41, 96
  impact resistance 31
  multi-axial strength 29
  permeability 35–36, 52
  resistivity 36
  shear strength 29, 50–55
  shrinkage 33, 69
  stress: strain curves 31–32
  tensile strength 28–29, 56–58
  thermal conductivity 19–20, 38–39,
    116, 127, 130
  thermal expansion 19–20, 37, 69–72
  water absorption 35
pumping
  admixtures 88
  Canary Wharf trials 88–91
  equipment 86–92
  water absorption 80, 86
raw marerials
  blastfurnace slag 4, 7–8, 21
  clay 4, 8–10
  pulverised fuel ash 10–12, 152
  shale 12–14
  slate 7, 12–14

Raymond Hillard Centre   117
Redesdale bridge   162
refurbishment
   bridges   166
   buildings   141–142
reinforcement detailing   64–66, 72–73,
   156–157
resistivity   36
Roxburgh County Offices   115
Royal Exchange   142
Russlite   4, 17
Russian standards   20

Shalite   132
shear of beams and slabs   55–60, 63, 67–
   68, 72
Sheraton Park Tower   124
ships
   American   17, 170–173
   European   173–174
   floating dock   175–176
   LNG tanker   173–175
ski-jumping platform   180
slip forming   111–112, 117
Solite   4, 12
Standard Bank, Johannesburg   122–123

steam curing 170
strength: density ratio   29–31, 44–45,
   109–110
Student Union, San José   130

thermal conductivity   19–20, 38–39
thermal expansion   19–20, 37, 69–71
Torre Picasso   125
torsion   60–61
transport   85–86
Twickenham grandstand   187–189

Ulft bridge   162–164
USS Selma   172–173

vacuum de-watering   97
vacuum soaking   79–80, 90–92

water
   absorption by aggregates   21–23,
      78–80
   absorption by concrete   35
   addition on site   85–86, 92, 104–105,
      111, 165

yield   85–86

Printed and bound by CPI Group (UK) Ltd, Croydon, CR0 4YY

01/11/2024

01782605-0009